Train Master

The Most Useful Locomotive Ever Built

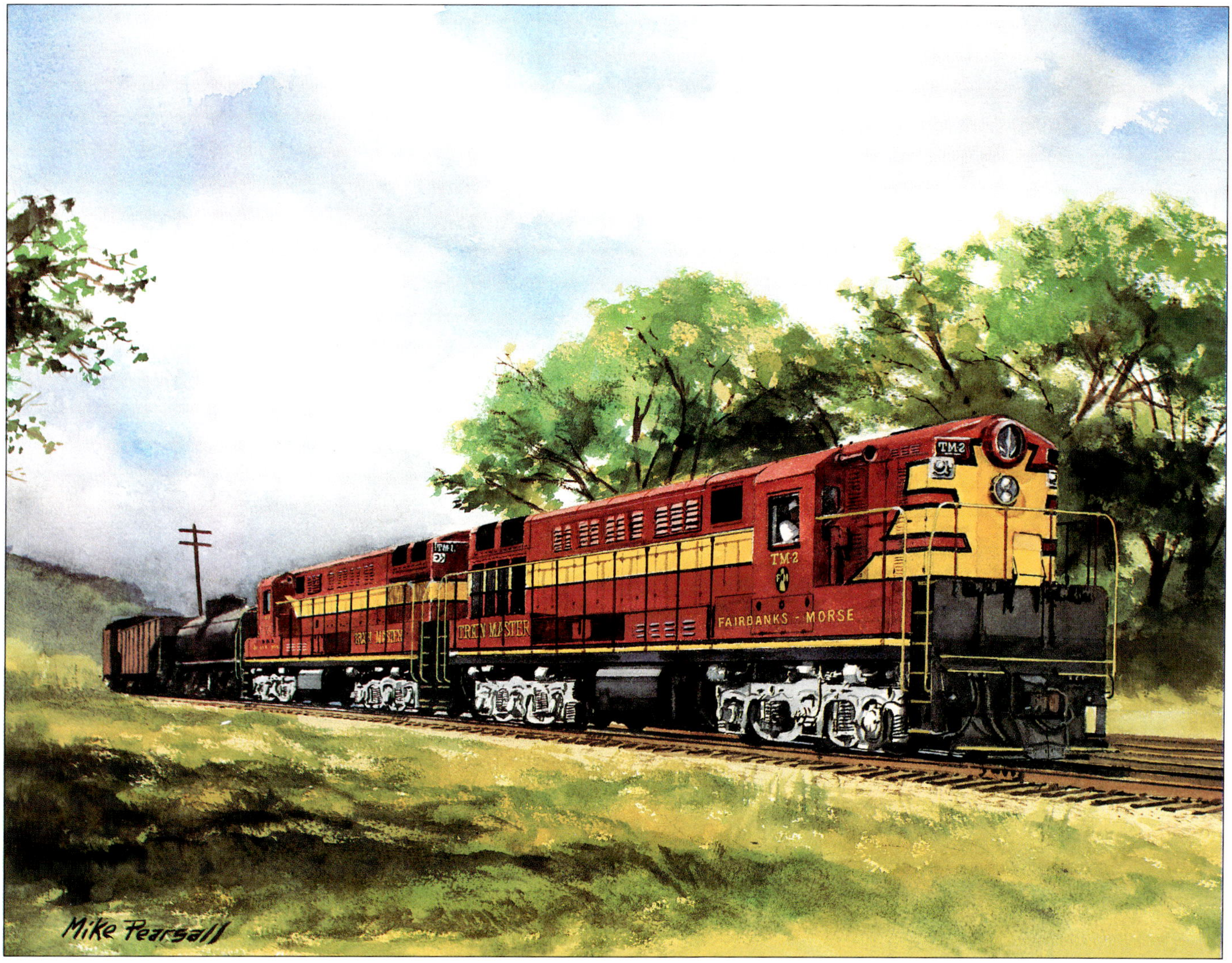

by *Diesel Era*
with David R. Sweetland

528 Dunkle School Road, Halifax, Pennsylvania 17032

Acknowledgements

Sincere thanks to the following photographers or collectors whose work appears here or was made available: Richard O. Adams, R.C. Anderson, Kenneth M. Ardinger, Keith E. Ardinger, Samuel Caliciotti, Warren L. Calloway, Jim Claflin, Frank DiFalco, Ken Douglas, George Drake, J.D. Eisenhart, Jr., Dave Engman, William Folsom, J.D. Ingles, Peter Gary, J. Michael Gruber/Mainline Photos, Emery Gulash, John D. Hahn, Jr., W.L. Hammond, Herbert H. Harwood, James C. Herold, Al Holtz, James Hope, Ed Kanak, Allen Keller, E. B. Kelsey, Robert Krone, Hans Kuring, Bill Linley, Gordon Lloyd, Jr., Dave Lustig, Kenneth MacDonald, G. M. McDonald, Louis A. Marre, William R. Martin, Tom Marsh and Overland Models, M. D. McCarter, George Melvin, Bruce R. Meyer, Gordon B. Mott, Gary Oliver, R.D. Patton, Tom Peebles, Bob Pennisi-Railroad Avenue Enterprises, Ralph L. Phillips, Don Pope, R. Post, J.R. Quinn, William Raia, Larry G. Russell, Dennis Schmidt, C.N. Shankweiler, F. D. Shaw, Jim Shaw, Richard Short, Fred Stephens, John F. Sullivan, William A. Swartz, Alvon Thoman, F.W. Trittenbach, R.R. Wallin, Robert Watson, Russell Wilcox, J. Harlen Wilson, Robert F. Wilt, Jim Wozniczka, and Martin S. Zak.

Also, a special thanks to: R. F. Corley for supplying information on the Canadian TMs; Greg Dickinson and Sam Caliciotti on the Lackawanna TMs; Joseph A. Strapac on Southern Pacific's TMs; and Robert M. Stacy, a former Fairbanks-Morse engineer who furnished photographs and technical information about Train Masters from his files.

Additional thanks to Brad Lomazzi of Western Railroad Collectables for supplying the various FM manuals and brochures that appear in this book; and Preston W. Cook and Don Dover of *Extra 2200 South* for making available the drawings that appear in this book. ★

Max Tschumi/Larry G. Russell collection

A trio of Canadian Pacific Train Masters is bathed in the floodlights of the Coquitlam, British Columbia, engine facility on October 23, 1962, awaiting the call to service.

Cover and title page - *Fresh from the paint shop and attired in their distinctive demonstrator paint scheme, TM-1 and TM-2 depart the Fairbanks-Morse plant at Beloit, Wis., for their debut at the Railway Supply Manufacturers' Association convention in Atlantic City, N.J., in June 1953. Artist Mike Pearsall has captured the colorful scene in an image that was inspired by a black-and-white photograph supplied by Robert M. Stacy.*

Produced by the Withers Publishing staff:
Layout and design – Paul K. Withers
Editor – Dan Cupper

First Edition, First Printing – March 1997
Typesetting by Withers Publishing, Halifax, Pennsylvania
Printed and bound by Paulhamus Litho, Montoursville, Pennsylvania
Manufactured entirely in the United States of America
International Standard Book Number: 1-881411-13-3
Library of Congress Catalog Card Catalog: 97-060477
Copyright © 1997 by Diesel Era

Withers Publishing
528 Dunkle School Road
Halifax, Pennsylvania 17032-9424

Alvon Thoman

Displaying the three paint schemes worn by Southern Pacific Train Masters are 4811, the only H24-66 painted in this simplified black and orange scheme; 4807, in the classic black, silver, orange, and scarlet "black widow" scheme; and 4813, wearing the gray-with-scarlet-wings scheme. Mission Bay roundhouse, San Francisco, February 27, 1959.

Table of Contents

Alvon Thoman

Southern Pacific Train Masters 4803 and 4802 wait patiently at 5th and Townsend Streets in San Francisco on November 23, 1957. The pair is powering a Stanford-Cal Big Game Special to Palo Alto, Calif. SP 4802-4809 were built in December 1953 under FM contract LD-164A, with half the group coming from a canceled New York Central order.

Richard O. Adams

One of two Reading Train Masters painted into the carrier's yellow and green scheme introduced with GP30s delivered in 1962, 265 and GP30 3609 lead a westbound coal train at Bethlehem, Pa., on October 7, 1967. Both units wear their second Reading road number – 265 was built as 867 and renumbered in 1967, while 3609 was built as 5509 and renumbered in 1963-64.

"The most useful locomotive ever built."

– Slogan used by Fairbanks, Morse and Co. at Railway Supply Manufacturers' Association convention, Atlantic City, N.J., in June 1953, to promote its then-new H24-66 Train Master road switcher.

From the start, the Train Master diesel locomotive was a curious creation. Coming from the smallest of the serious diesel locomotive builders, it was the most powerful single-engine road-switcher unit in the diesel-electric locomotive market. But it was more than that. Its builder, Fairbanks, Morse & Co., intended it to be a universal workhorse, one that, through the use of a skillful advertising campaign, promised to be all things to all railroads.

Although this was a tall order for a new model, the FM Train Master lived up to its billing, but on a much smaller scale than that for which its designers had hoped.

A Brief History

In the early 1930s, FM began developing a medium-speed, heavy-duty, minimum-weight diesel engine for use in services that required maximum power in a minimum of space. The result was the opposed-piston diesel engine, which, in effect, combined two in-line engines by inverting one on top of the other and joining pairs of cylinders into a common combustion chamber. By doing so, each cylinder contained two pistons, each moving toward the center of the cylinder. The head structure was eliminated, with the pistons forming the combustion space between the facing crowns. Fuel was injected in the center of the cylinder length.[1]

The design also eliminated valves and valve operating assemblies. Exhaust and air intake were handled through two sets of ports cut into the cylinder walls. As the pistons are pushed apart by the force of combustion, they uncover these ports, permitting exhaust gases to escape through the bottom ports, while fresh air charges the cylinder through the top ports. The movement of the pistons is timed so that the exhaust ports open first, and the air ports close last, assuring complete scavenging or removal of the burned gases from the cylinder.[2]

Light and extremely compact, this design was well-suited for use in loco-

J. D. Eisenhart, Jr. collection

The centerpiece of Fairbanks-Morse's Atlantic City, N.J., 1953 trade-show display was the pair of colorfully painted Train Master demonstrators, TM-1 and TM-2. Note the extra tri-mount truck for potential customers to inspect.

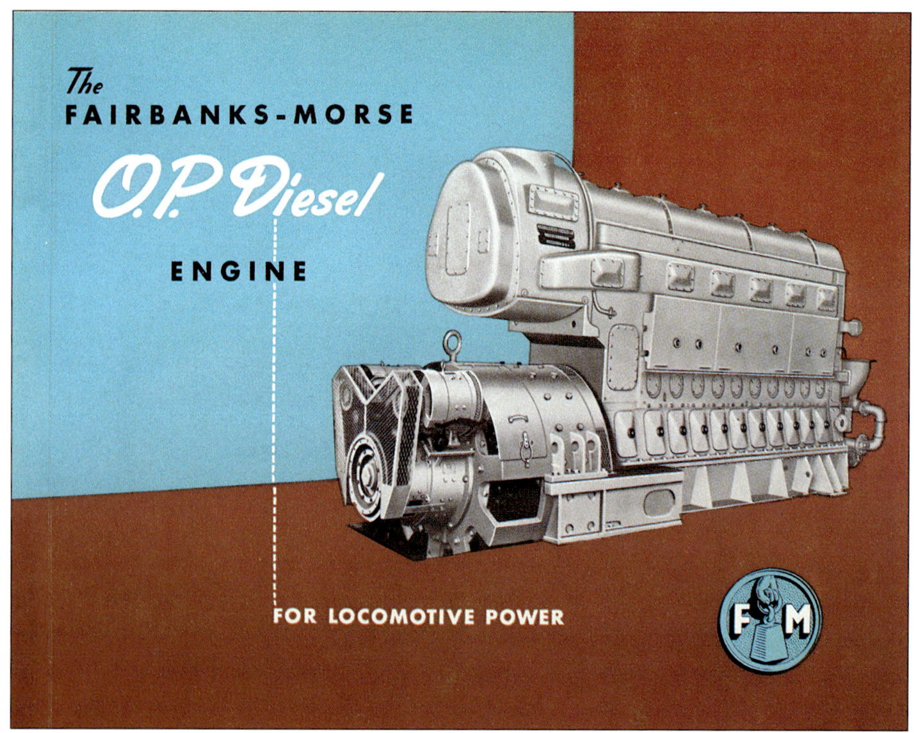

courtesy of Western Railroad Collectibles

To promote the Model 38 opposed-piston diesel engine, FM produced this slick brochure in 1948. Containing overlaying cutaway views, the booklet extolled the merits of the opposed-piston design coupled with the two-cycle combustion principle.

motive carbodies. The first railroad use of this engine occurred in 1939, when a group of motor cars was powered with a 750-horsepower, five-cylinder version of this power plant. With the coming of World War II, the U.S. Navy absorbed nearly all of FM's engine production and installed the power plants in a variety of combat ships. Eventually, more than half of the Navy's submarine fleet was equipped with FM's opposed-piston engine.

As World War II ended, Fairbanks-Morse again entered the railroad market, this time with a yard switcher design. Completed in August 1944, the first switcher, Milwaukee Road 1802, was powered by six-cylinder, 1,000-horsepower version of the 10-cylinder, 1,800-horsepower power plant used in submarine service.

Seeking to offer a suitable mainline diesel unit to America's railroads, which were quickly phasing out steam engines, FM in 1945 designed a locomotive that was powered by the same 10-cylinder engine used in submarine service, but reworked to deliver 2,000 horsepower. The result was the Erie-built model, so named because it was built at the Erie, Pa., plant of FM subcontractor

General Electric. Typical of many early diesel road locomotives, its carbody was a cab-unit style, with the operating cab at one end and the engine crew looking forward over a nose. Although it offered 500 more horsepower than the comparable EMD cab-unit model, the F3, the Erie-built proved to be costly to build, and met with limited market appeal.

American Locomotive Co.'s relative success with its road-switcher carbody design (with operating cab located near one end, between a short and a long hood), the RS series, and FM's foresight in wanting to offer a comparable model, brought two new diesel models to the nation's railroads in 1947. Using the same 10-cylinder engine that powered the Erie-built, FM's H20-44 was the first single-engine B-B diesel built with a hood-type carbody carrying a 2,000-horsepower rating. It was years ahead of its competition. For customers not needing the extra horsepower, FM offered a 1,500-horsepower model, the H15-44. Both models found limited acceptance, as a widespread market for road-switcher- or hood-style locomotive models had yet to develop. This did not discourage the company, but FM, in fact, retained the line and would add an upgraded version of the H15-44, the H16-44, in 1950. But in the late 1940s, the cab unit was still the model of choice for road locomotives among America's railroads; to this end, FM in late 1948 announced the introduction of the Consolidation line of cab units.

In 1950, FM delivered its first C-Line unit. Eight feet shorter than the Erie-built, it could support an eight-cylinder 1,600-horsepower, 10-cylinder 2,000-horsepower, or 12-cylinder 2,400-horsepower opposed-piston diesel engine. With a variety of options available, the unit could be used in either freight or passenger service. But FM delivered only 49 C-Liners in 1950, and only 22 the following year. What went wrong? The C-Liner was introduced five years too late. The cab design was falling from favor with the nation's railroads, which were changing their preferences to the all-purpose utilitarian hood unit (EMD's successful GP7 model was introduced in late 1949). Thus, in the fall of 1951, FM's Railroad Division requested funds to design yet another new locomotive model, a 2,400-horsepower six-axle hood (road-switcher) unit.

The Train Master

By April 1952, under the direction of FM designer George Derrig* (who had worked under Henry "Hank" Schmidt*, the manager of the C-Liner project),

Ken Douglas

Recognizing the limitations of the cab unit in secondary freight service where switching was performed en route, FM designed a road-switcher carbody to compete with Alco's RS series. Always the innovator, FM offered its road switcher model, the H20-44, with a 2,000-horsepower engine, 500 more horsepower than any model offered by its competitors. A bridge-line carrier, Pittsburgh & West Virginia Railway, chose the H20-44 to use in dieselizing its operations, acquiring 22 units between 1947 and 1953. Rook, Pa., February 3, 1963.

enough progress had been made on the design that the company's marketing group, headed by an FM newcomer, George S. Cohan, gave the model its name: Train Master.

Hired in March 1952, Cohan brought with him advertising expertise previously unknown at Fairbanks-Morse. To market the new locomotive, a series of 12 montages was created by Francis Chase to illustrate the multi-faceted uses of the Train Master. In addition to running them in a 12-part advertising campaign, which ended in June 1953 just as the first two demonstrators were arriving at the Atlantic City show, the company mailed a promotional brochure in late 1952 containing the same artwork. Along with the brochure, a note from V. H. Peterson** was enclosed that read: "So many people have expressed interest in the new Fairbanks-Morse Train Master locomotive, that we are sending our colorful TM book to a select list of friends who we feel should have a personal copy." This only served to reinforce the commonly used "get on the bandwagon" advertising approach.

Although the Train Master (which FM designated as its model H24-66 - Hood, 2,400 horsepower, six motors, six axles) was marketed as an all-new locomotive, it incorporated many design concepts carried over from previous FM locomotives, including the Consolidation line. The heart of the Train Master was the 12-cylinder, 2,400-horsepower, two-cycle opposed-piston power plant that had been offered as an option in the C-Line – a total of 22 2,400-horsepower C-Liners were built – and that had been available since 1951.

The most powerful single diesel engine in locomotive service, the engine incorporated the same basic working parts such as pistons, rods, bearings, liners, injection nozzles, and so forth, that were used on all FM opposed-piston engines in locomotive service. It differed from the other opposed-piston engines only in number of cylinders, length of crankshaft and overall dimensions directly related to its 2,400-horsepower rating. Most of the engine's earlier problems were cured with the use of aluminum bearings, counterweighted crankshafts, and explosion-resistant cylinder blocks that had been introduced with the Consolidation Line in 1950.

By 1953, reinforced pistons and

courtesy of Western Railroad Collectibles

The famous "Blue Brochure" introduced the Train Master with a series of montages created by Francis Chase that illustrated the locomotive's versatility.

To describe the Train Master's features, FM produced this brochure in September 1952. Its first section covered the power plant, transmission, weight, supplies, train heating, auxiliaries, and control features; the second part covered the air brakes and dynamic brakes; and the final part covered dimensions, running gear, carbody construction, and miscellaneous equipment.

improved chrome cylinder liners were in use, further improving the opposed-piston engine's reliability. To maintain the critical alignment between the motor and generator, these two components shared a common sub-base that maintained alignment regardless of underframe deflection (temporary dis-tortion from its original shape). As with the C-Line, several key engine support systems were modular or unitized in design. The air compressor and cooling system, as well as optional dynamic braking and passenger heating boiler, were designed for ease of maintenance and changeout. Improved and upgraded AC traction motor blowers and AC radiator cooling fans from the C-Line were used in the Train Master, instead of the DC accessory equipment used on previous FM road switchers.

As initially offered, the main generator in the Train Master was a Westinghouse Model 498BZ, an upgraded version of the Model 498B, the same generator found in the 2,000- and 2,400-horsepower C-Line models. When Westinghouse announced that it was departing the heavy traction equipment business in 1953 (final deliveries were made in June 1954), FM turned to General Electric to supply electrical gear for its locomotives. A GE Model GT567-series main generator replaced the Westinghouse equipment.

Likewise, Westinghouse Model 370 traction motors, found on a majority of C-Line locomotives, and used under early Train Masters, were replaced with GE Model 752-series motors. With the change in traction motor manufacturer came a change in wheel diameter, from 42-inch wheels on Westinghouse-equipped units to 40-inch wheels on the GE-equipped units (although Wabash and Jersey Central ordered Train Masters with GE equipment and 42-inch-diameter wheels). After the switch to GE equipment was made, Westinghouse equipment was offered as an option for as long as it remained available.

One notable difference between the Train Master and previous FM models was the all-new truck design. Although FM's 1951-built H16-66 road switcher rode on six-axle trucks, they featured rigid cast frames, a design that was unsuitable for the high-speed service FM designers had in mind for the Train Master. As a result, FM engineer Jim Long and an industry supplier, General Steel Castings, worked to design a new tri-mount truck. Similar to the design used on Alco's RSD-model road switchers, the tri-mount truck employed a three-point loading principle, but offered several design changes to reduce maintenance costs as well as improve maintenance access.

The tri-mount truck used the principle of three-point loading. This method had long been used successfully as the loading principle on high-speed electric locomotive running gear on nearly every

Covering both types of electrical equipment, Fairbanks-Morse produced Operator's Trouble Shooting Manuals. These 52- to 56-page booklets were issued to provide "readily accessible and concise information on minor and easily remedied operational faults that might occur on-the-road."

Trains collection

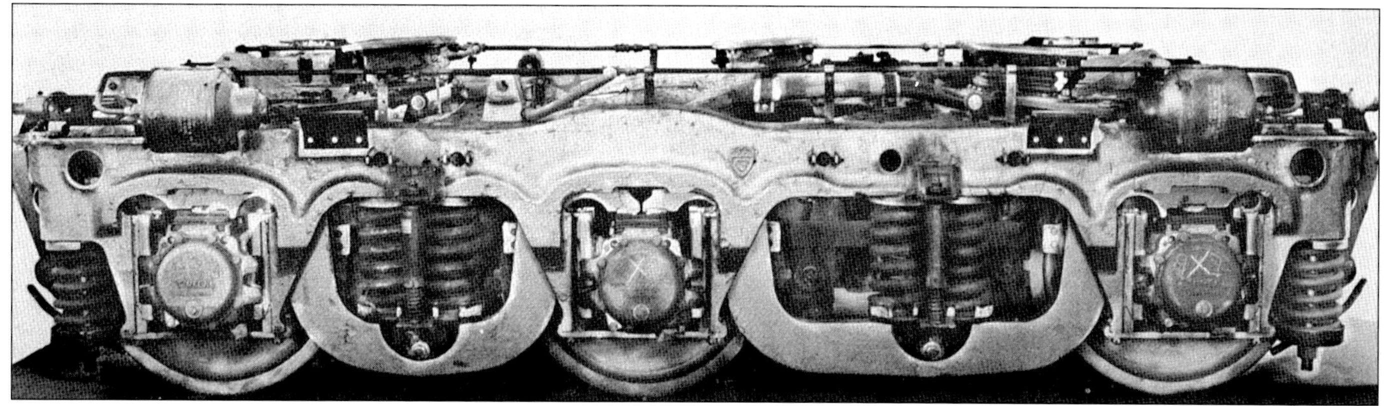

While the opposed-piston engine was unique among diesel designs, the truck style FM employed on the Train Master was equally intriguing. Equipped with 40- or 42-inch wheels, the three-motor truck boasted a wheelbase measuring 156 inches.

electrified railroad in North America, and offered a truck with superior tracking control. On the FM truck, the center plate was mounted on the Number 1 transom while two mounting pads were mounted near the outer ends of the Number 2 transom. By spreading the weight of the locomotive along the entire length of the frame, the frame was stabilized and truck frame reaction to track irregularities was reduced.

Unlike other truck designs with the center plate directly above the center traction motor, the Train Master tri-mount truck featured an off-center center plate that allowed easier access to the center traction motor. As a result, all traction motors were equally accessible for inspection and servicing. Unlike Alco's tri-mount truck, which is double-equalized, the FM design was single-equalized. This allowed the truck to take full advantage of the six-wheel design's known ability to absorb high and low spots in the track without sacrificing accessibility. The use of extended single equalizer bars gave added clearance required for inspection and servicing of brake rigging and springs, a complaint commonly heard from mechanical personnel servicing Alco's double-equalized truck design.

With these truck improvements, FM advertising brochures touted the Train Master as being able to deliver "five percent more tractive effort than a conventional four-axle locomotive where adhesion is limiting and has delivered a tractive effort equal to 32 percent of the locomotive's total weight on drivers." This was at a time when most railroads used a 25 percent figure when calculating starting tractive force.

The Train Master was a massive locomotive – 66 feet long and 15 feet

courtesy of Western Railroad Collectibles

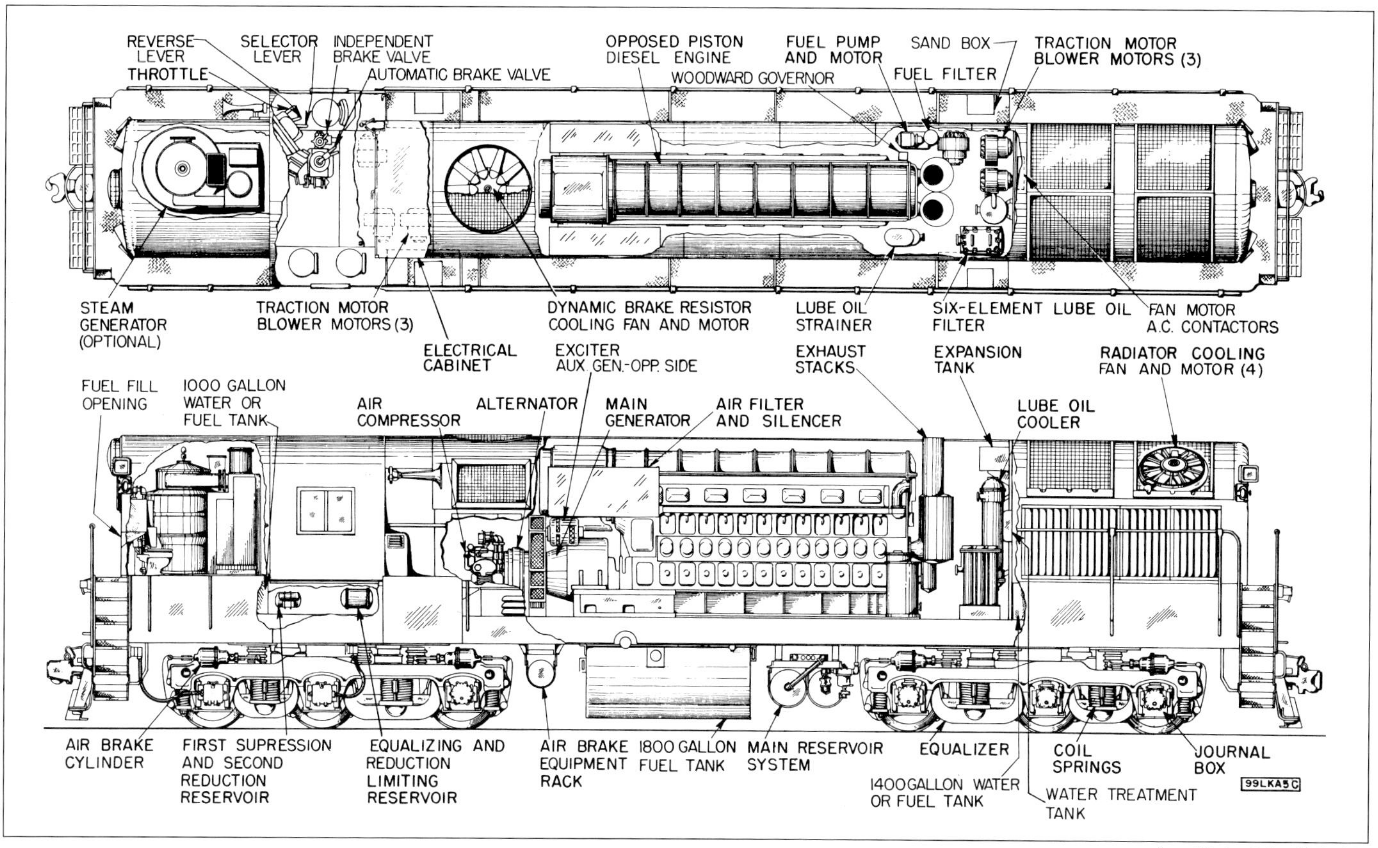

From Bulletin 1602G, November 1956, Engineman's Manual for Operating Fairbanks-Morse 2400 HP Train Master Locomotives with General Electric Rotating and Control Equipment with Woodward Governor and Static Excitation for the Delaware, Lackawanna and Western Railroad (road numbers 860 and 861, FM serial numbers 24-L-1035 and 1036).

courtesy of Western Railroad Collectibles

Bulletin 1602A, June 1955, for the Southern Railway System (road numbers 6300-6304; FM serial numbers 24-L-856-860).

courtesy of Western Railroad Collectibles

Bulletin 1302, August 1953, for 1,600-horsepower general service and 2,400-horsepower Train Masters.

high – but was light on its feet. It weighed 343,000 pounds; with steam generator, including water supply, and dynamic brake, the unit's overall weight increased to 366,000 pounds. Up to 9,000 pounds of ballast could be added, bringing the maximum weight to 375,000 pounds, although additional ballast could be added when extremely high tractive force was desired (in fact, Virginian units 69-74 weighed 396,600 pounds, the heaviest H24-66s built). FM promoted the fact that the Train Master provided 50 to 60 percent more power, 50 to 90 percent more continuous tractive effort rating, and 50 percent more weight on its wheels in a single unit as compared to any 1,500- or 1,600-horsepower four-motor, four-axle diesel locomotive then in service. The Train Master equaled one and a half H16-44s, and two Train Masters could replace three 1,600-horsepower B-B units.

The basic Train Master was equipped with an 1,800-gallon fuel tank with two optional tanks, a 1,000-gallon tank located in the floor of the short hood, and a 1,400-gallon tank on the rear end platform. These tanks could be used either for fuel or for train heating water. Sand capacity was equally impressive. Four sand boxes, each with a capacity of 24 cubic feet, supplied sand to the six-axle trucks. When compared to the typical 1,500/1,600-horsepower road switcher of the era, the TM carried 50 percent more fuel, from 50 to 100 percent more water, and from 60 to 200 percent more sand.

FM offered three gear ratios for the Train Master with Westinghouse traction motors: 68:15 for 65 mph (maximum speed), 63:15 for 70 mph, and 62:17 for 80 mph operation. With General Electric motors, gear ratios of 65:18 and 74:18 were available.

With all these superlatives, it was no wonder that the FM advertising department chose the name Train Master for such a revolutionary locomotive. FM's promotional literature stated that the model was developed "to meet the increasing demand for a practical motive power unit embodying high engine horsepower and high pulling capacity. Each of the conventional diesel locomotive models now used, although highly flexible when compared to steam locomotives, is limited to certain classes of service by its power, transmission, or general design. The Train Master includes the horsepower, transmission capacity, weight on drivers and general design features to fit it for any class of service. Furthermore, and of great importance to the railroads, all of the appurtenances and accessories required for all classes of service can be applied to any one unit, and it is not necessary to substitute one feature for another." It was indeed, "the most useful locomotive ever built."

Production Begins

The target date for completion of the first TM was set for the week of June 22, 1953, in order to be displayed at the Railway Supply Manufacturers Association exhibit at Atlantic City, which would be attended by some 9,075 representatives of the industry, both buyers and suppliers. By mid-April, the first Train Master (TM-1) left the company's Beloit, Wis., plant and instead of breaking in on FM's usual "test track," Milwaukee Road's RSW Division (later Milwaukee Division) east of Beloit to Sturtevant, it tested west of Beloit to Freeport, where several miles of highway paralleled the railroad – permitting FM personnel to pace their new creation and film the newly designed truck action at track speed. Painted solid orange and riding on silver-painted trucks, TM-1 was joined by TM-2 in early May. The tests proved very successful and the units were returned to Beloit for repainting into a colorful red and yellow scheme with silver-painted trucks and black underframe for the trip east. Departing Beloit on May 16, 1953, the pair traveled via Milwaukee Road to Chicago, where they were publicly displayed for the first time at Union Station. After that, they were turned over to the Pennsylvania Railroad.

Their first assignment on the Pennsy was handling eastbound Train 44, a Chicago-to-Pittsburgh mail and express run. In addition to testing on the PRR, TM-1 and TM-2 were tested prior to the Atlantic City show by Detroit, Toledo & Ironton, Baltimore & Ohio, and Western Maryland.

A second pair of demonstrators was released by Fairbanks-Morse at the end of May. "Western" demonstrators TM-3 and TM-4, also carrying the red and yellow demonstrator scheme, were bound for service on Rock Island, Chicago & North Western, Missabe Road, and Illinois Central. After moving further west, TM-3 and TM-4 worked on Union Pacific, Southern Pacific, Rio Grande, Santa Fe, and Great Northern.

After the trade show closed, TM-1 and TM-2 tested on Reading, New York Central, Central Railroad of New Jersey, New Haven, Long Island, Virginian, and Wabash, as well as putting in a second stint on the Pennsy. Equipped with steam generators, they were tried in a variety of freight and passenger assignments. For passenger train heating, the Train Master offered a 4,500-pound-per-hour steam generator and 2,400 gallons of water storage, adequate for heating 15 passenger cars in zero-degree weather. Interestingly, TM-4 lacked this equipment. Over the eight-month demonstration period, the four TMs covered 170,000 miles on 20 railroads.

Production

Long before the Atlantic City exhibit, Fairbanks-Morse had snagged its first H24-66 customer. In November 1952, Delaware, Lackawanna & Western ordered 10 Train Masters, at $250,000 each, to complete its dieselization. Lackawanna's Train Masters, 850-859, equipped with dynamic brakes and steam generators, were the carrier's first road switchers to wear the attractive maroon and gray color scheme previously found only on mainline passenger and freight cab units. They were also Lackawanna's second Fairbanks-Morse order, and represented something of a coup on a railroad that had previously been almost exclusively been committed to EMD power, but also was very close to ending steam operations. Delivered in June 1953 with a 68:15 gear ratio (65 mph) for both passenger and freight service, the big Train Masters ran with the long hood designated as front.

Following the Lackawanna order, Reading Co. took delivery of four Train

Both TM-1 and TM-2 demonstrated on the New Haven during August 1953. In this view at Hartford, Conn., on August 9, 1953, TM-1 was operating in passenger service between Springfield, Mass., and New Haven, Conn. Although both units performed well and no orders were forthcoming, this was probably a courtesy call, as NH had purchased 10 H16-44s in 1950 and 10 2,400-horsepower C-Liners delivered in 1951-52. Additional FMs arrived three years later in the form of 15 additional H16-44s.

David R. Sweetland

Masters, two for freight service, numbered 800 and 801, and two for passenger service, 860 and 861. Delivered in October 1953, they were followed by a second order for five freight-service units, numbers 802-806. These nine H24-66s replaced 1,600-horsepower units on a two-for-one basis and accelerated dieselization of the Reading. All units had Westinghouse electrical equipment and operated with the short hood designated as front.

In 1953, NYC placed an order for eight Train Masters, assigned numbers

The Train Master's specifications were impressive enough to persuade one customer, Delaware, Lackawanna & Western, to place an order even before the first two demonstrators were released for service. In July 1957, DL&W 858 easily propels a four-car westbound commuter train at Passaic, N.J. The dual-service TMs worked the daily commuter runs from Hoboken, N.J., as well as handling freight assignments.

Al Holtz

Built for passenger service, Reading 861 was one of four Train Masters delivered to the carrier in the fall of 1953. Two of the units, including 861, were equipped for passenger service and carried steam generators in their short hoods. Although the steam generator vent is intact, the equipment would not be needed this day, as the unit had drawn a way freight assignment when it was photographed in the Philadelphia area in September 1962.

4600-4607, for operation out of Ashtabula, Ohio, in coal train service, but the order was later canceled. It is not known if the cancellation was caused by a change in railroad management, or by operating-department second thoughts after Central experienced main generator flashovers with its 4500-series C-Liners at high speed. Both the CPA24-5 C-Liner and the Train Master used the same Westinghouse 498 main generator. On 2,400-horsepower locomotives, it was shunt circuited at 1,080 volts which made commutation difficult to stabilize, and any disturbances – rough track or wheel slip – could cause destructive flashovers[2] (something like having the main generator struck by lightning [3]).

After a series of meetings in the fall of 1953, Southern Pacific purchased 16 Train Masters, including TM-3 and TM-4, which had successfully demonstrated on Espee property. TM-3 and TM-4 became SP 4800-4801 before the end of 1953, and 4802-4815 were assembled between December 1953 and February 1954 at Beloit. Eight of the SP units came from the canceled NYC order.

Virginian followed right behind SP with an order for 19 Train Masters, the largest Train Master order FM had received to date. Unlike previous purchasers, which had experience with diesel-electrics, Virginian was still 100 percent steam- and straight-electric-powered. The carrier chose to replace steam with two types of diesels, 2,400-horsepower Train Masters for the western end of the system, where they replaced heavy articulated steam locomotives, and 1,600-horsepower H16-44s for the eastern half of the system. Specifications for this order included the first use of General Electric electrical equipment. Completed in April and May 1954, Virginian 50-68 were painted yellow and black; the long hood was designated as front.

One more Train Master order completed H24-66 deliveries for the year 1954. Impressed with the demonstration tour of TM-1 and TM-2, Central Railroad of New Jersey ordered seven units, numbered 2401-2407. Equipped with steam generators, they were delivered with 65:18 gearing and 42-inch-diameter wheels. Assignments included mainline freight, passenger, and commuter service, commuter operation on the New York & Long Branch Railroad, and Reading pool service to Philadelphia and Potomac Yard south of Washington, D.C. The delivery of CNJ Train Masters 2401 and 2402 in May, and ending with 2407 in July 1954, effectively ended steam operations on Jersey

above, *Southern Pacific H24-66s 4809 and 4810 lead 89 cars of mixed freight through Fort Bliss, Texas, on December 5, 1954. The units, which had been on the roster for about a year, were working on the Rio Grande Division, but soon, the desert sand would create problems for the FM's air intake filters, causing them to be reassigned to California.*

left, *Posing for its official portrait outside FM's Beloit, Wis., plant in April 1954 , Virginian Train Master 58 would soon be tying on to endless strings of hopper cars, replacing aging articulated steam engines.*

Central. Shortly after delivery, the carrier experienced vibration problems with the tri-mount truck at high speed. Reading also experienced these problems, and after a series of tests on Reading, both roads' H24-66s were returned to Beloit, where a modified truck bolster system was installed.

After nearly a year's gap in production, Train Masters again began rolling down the Beloit assembly line. In May 1955, the first unit of a five-unit TM order was completed for a new customer, Southern Railway. Assigned to its Cincinnati, New Orleans & Texas Pacific subsidiary, numbers 6300-6304 were delivered wearing the carrier's green and imitation aluminum (white) paint scheme. In keeping with their intended role, the units were equipped with 74:18 gearing and built with low end platforms for operation with other manufacturers' models. They were run between EMD F7s on Southern's "Rat Hole" division between Cincinnati and Chattanooga, Tenn.

FM didn't overlook the Canadian market, for which it had built earlier models. Beloit released Canadian Pacific 8900 in June 1955, and Canadian National 3000 a month later. Both were shipped from Beloit to the company's Canadian Locomotive Co. plant at Kingston, Ontario, where finishing touches were applied. Wearing both CLC and FM builder's plates, the units then tested on their respective railroads. CP 8900 entered service on July 13, 1955, and CN 3000 was placed into service in August.

CN 3000's first assignment was to the Manitoba District where, after a year of operation, it was renumbered 2900. In the early 1960s, it was assigned to the Toronto commuter pool.

The only other Train Masters built in 1955 were six additional units for Reading, the carrier's third order. Built with steam generators, dynamic brakes, and dual controls, units 862-865 were initially assigned to St. Clair, Pa., where they replaced I9- and I10-class 2-8-0s. For all intents and purposes, this group of units ended steam operations on the anthracite carrier, although several members of Reading's famed 4-8-4 T-1 class returned to service in 1956 to handle a temporary traffic surge.

Robert Krone

With westbound Train 199, the Queen of the Valley, Jersey Central H24-66 2406 accelerates out of Elizabeth, N.J., on August 27, 1957, with a six-car consist bound for Allentown and Harrisburg, Pa. The unit wears CNJ's standard green paint scheme with two yellow stripes and the carrier's Statue of Liberty herald on the ends and cab sides.

FM/William Folsom collection

Wearing Southern's classic green color scheme, H24-66 6304 poses at FM's erecting shop in June 1955. Southern's five Train Masters were the largest diesel locomotives on the road until 1959, when EMD delivered SD24s.

Dan Cupper collection

R. R. Wallin

FM's first Train Master, TM-1, poses in its new owner's colors at Decatur, Ill., in 1957. Purchased (along with TM-2) by Wabash in 1954, the units were joined by six new Train Masters in 1956.

The first units built in 1956 were for Jersey Central, which had placed a second order for H24-66s. With the carrier's fleet of double-ended Baldwin passenger units reaching the end of its economic usefulness, CNJ ordered six H24-66s for delivery in March 1956. The units replaced their Baldwin counterparts in commuter service.

Impressed with their demonstration tours, Wabash had acquired two of the demonstrator units, TM-1 and TM-2, in February 1954. Renumbered 550 and 551, they were then joined by six built-new Train Masters, 552-554 and 552A-554A, in mid-1956. Unlike the demonstrators, these units were built with low end platforms. These six also lacked dynamic brakes, reflecting the carrier's essentially table-top gradient profile. TM-1 and TM-2 were originally equipped with this option, but prior to their delivery to Wabash, it was removed. Unlike 550 and 551, which were built with Westinghouse electric equipment, the rest were built with General Electric electrical gear. To maintain uniformity, these units were built with 42-inch-diameter wheelsets like the demonstrators. Assigned to Decatur, Ill., for maintenance, these were the only six-motor units ever purchased by Wabash.

Pennsylvania Railroad became the 10th Train Master customer, ordering nine units numbered 8699-8707. Delivered in August and September 1956, they carried PRR classification FS-24M and were initially assigned to hump and transfer service at Columbus, Ohio, and helper service out of Pitcairn, Pa., on the West Slope grade.

North of the border, the successful operation of Canadian Pacific 8900 during 1955 resulted in a 20-unit order for 1956 delivery. These units were assigned to the western half of the system, with units initially assigned to Calgary, Alberta, for Calgary-Vancouver service and Nelson, British Columbia. To replace 1,600-horsepower steam-generator-equipped units in winter passenger service (on a three-for-one basis), four Train Masters came equipped with two steam generators. To accommodate this equipment, CP 8901-8904 were delivered with full-width short hoods. These CLC-assembled units were built between June and October 1956 at Kingston, Ontario, and were delivered with the long hood designated as front.

At FM's Beloit factory, a second Lackawanna order for H24-66s was filled when a pair of units was completed in November 1956. Although they displayed passenger-series numbers, 860 and 861, they were not equipped with steam gen-

J. David Ingles/Louis A. Marre collection

The six new Train Masters purchased by Wabash were built with low end platforms and lacked the open space under the side walkways found on the demonstrators. Jacksonville, Ill., May 2, 1962.

David R. Sweetland

The giant Pennsylvania Railroad purchased only nine Train Masters, but had approached FM about a 50-unit order that never materialized. One of PRR's two sets of H24-66s that were assigned to helper service awaits the call to duty at Conemaugh, Pa., in March 1961.

erators. Carrying GE electrical equipment, these H24-66s were assigned to freight service and plied the Buffalo-to-Hoboken main line, typically in 4,800-horsepower sets indiscriminately mated to their steam-generator-equipped counterparts.

Just before Christmas 1956, the Reading Co. acquired its last two Train Masters, 807 and 808. Unlike previous Reading H24-66 orders, these two units were built with GE electrical equipment and together with other freight-service Train Masters, 800-806, worked out of Reading's Harrisburg-area yard at Rutherford, Pa. These eight units, along with PRR's nine Train Masters, were the only H24-66s to carry 4,200 gallons of fuel, the largest capacity available on the TM.

The final year in which FM built Train Masters was 1957, with only six units assembled for Virginian Railway. Units 69-74, shipped from Beloit in May and June 1957, were similar to locomo-

William A. Swartz/M. D. McCarter collection

The Train Master found limited acceptance north of the border. Only Canadian Pacific purchased the model in quantity, taking delivery of 20 units in 1956. Four of those were equipped with two steam generators each, which required a full-width short hood, as seen on 8902 at Medicine Hat, Alberta, on October 15, 1956.

Richard Short/David R. Sweetland collection

In late December 1956, Reading received its last two Train Masters, 807 and 808. The only two TM's on the Reading with GE electrical equipment, they were equipped with dynamic brakes and dual cab controls. Philadelphia, March 1963.

The next-to-the-last Train Master built, Virginian 73 is at Mullins, W. Va., on June 1, 1958.

tives in the road's first order, in that they were equipped with 1,800-gallon fuel tanks, but carried extra ballast, making them the heaviest Train Masters built, weighing 396,600 pounds. When Virginian 74 was fired up at Beloit in June, it signaled the end of Train Master production. In all, 127 units were built for 10 U.S. and Canadian railroads between 1953 and 1957. Virginian owned the most units, with 25, and Canadian National had the least, with only one.

What Might Have Been

Several missed opportunities could have placed the Train Master roster at twice its size and set the stage for a proposed supercharged Train Master rated at 3,600 horsepower (with only a 4,000-pound weight increase, virtually no increase in overall engine size or in lube oil or water consumption, and at a fuel saving of 5 to 10 percent at full load.[4]

First, Illinois Central's successful demonstration of the Train Master in October 1953 set the stage for what the builder thought would be a 75-unit order, but internal corporate problems at FM scared off several potential customers.

In 1955, feuding in the Morse family left the company vunerable to a corporate takeover which happened the following year, when Leopold Silberstein's Penn-Texas Co. obtained control of Fairbanks-Morse. Penn-Texas, a holding company, was then subject to its own internal management wrangling as Silberstein was ousted by Alfons Landa in 1957. Landa was in turn ejected in a stockholder-instigated management coup. Penn-Texas then changed its name to Fairbanks-Whitney. (by the early 1960s, the name had again been changed to Colt Industries, the historic name of another Penn-Texas subsidiary).

In 1957, FM was one of three locomotive builders that were asked for bids to complete dieselization of the Pennsylvania Railroad. Pennsy was interested in up to 50 Train Masters, but ended up placing orders only with EMD for 25 SD9s and Alco for 25 DL702/RSD-12s.

About a year before the demonstration of TM-1 and TM-2 in July and August 1953 on the New Haven Railroad, FM produced an artist's drawing showing proposed New Haven Train Masters 2401 and 2402 in green paint with gold striping, and operating short hood forward. NH did not place an order for Train Masters, but it did purchase 15 H16-44s for delivery in 1956.

It can be said that the Train Master started the horsepower race between locomotive builders. It would take Alco until late 1955 to introduce a comparable model, the DL-600A (RSD-7) with its 2,400-horsepower 16-cylinder 251-series engine, and EMD until 1959, with its turbocharged SD24 model, to match what upstart Fairbanks-Morse accomplished in 1953. Sales of only 127 Train Masters seemed small in comparison to 299 of FM's smaller H16-44 hood units, but it helped establish FM as the number-three diesel locomotive builder in the early 1950s, edging aside Baldwin – which had been, over the preceding century, North America's oldest and biggest locomotive manufacturer.

The downfall of this revolutionary model can be traced to several factors, the major one being opposed-piston engine problems experienced by early FM customers. Piston failures and crankcase explosions were commonplace with the builder's Erie-built model. This tarnished the reputation of the opposed-piston design in the eyes of some of the country's major railroad mechanical departments. A look at the customers for this model – Union Pacific, Santa Fe, New York Central, and Pennsylvania – reveals that the first three never purchased the TM, and the PRR sampled only a handful. Without these influential customers, did the TM ever have a chance for success? Although FM made major improvements in the OP engine with the introduction of the C-Line in 1950, problems persisted – a longtime FM customer, New York Central, began re-engining 2,000- and 2,400-horsepower C-Liners after only four years of service.

Another factor was the Train Master's initial purchase price. The average price of a TM was $260,000, or $108 per horsepower, while EMD's GP9 sold for $160,000, or $91 per horsepower.

While related to the early piston problems, FM's locomotive maintenance costs were typically the highest when compared to the other major builders. This is the direct result of the OP engine design, which required the upper crankshaft to be removed to do any work on the cylinders, including relining them, a common practice on earlier FM models. To avoid this labor-intensive operation, FM in 1954 began an engine exchange program. Stock engines were purchased from railroads that had either repowered their FM locomotives or scrapped them. Although contracting locomotive maintenance to an outside source is commonplace in the 1990s, it was unheard of in the 1950s. Railroads were still well staffed with maintenance personnel remaining from the steam era, an era when all repair work was done in-house.

For its day, the Train Master was a large diesel locomotive whose 2,400 horsepower and quick acceleration was years ahead of the competition. By the time the opposed-piston engine had become a reliable power plant, steam locomotives had already been vanquished from North America's rails, leaving no major customers for "The most useful locomotive ever built." ★

1. *Locomotive Cyclopedia, 15th Edition,* 1956, Simmons-Boardman, New York, N.Y., p. 146

2. Cuisinier, Win, Fairbanks-Morse H24-66 Train Master, *Extra 2200 South,* Issue 61, July-August-September 1977, p. 15.

3. Aldag, Robert Jr., How Fairbanks-Morse Got into the Locomotive Business, Part 1, *Trains,* March 1987, p.36.

4. *Trains,* May 1958, p. 57

* George Derrig and Henry "Hank" Schmidt were locomotive design engineers at EMD before being hired by FM in 1947.

** V.H. Peterson had been a locomotive salesman for Baldwin. He was hired by FM in 1947.

Phase Ia - *Southern Pacific 3020, built as FM demonstrator TM-3, illustrates the Phase Ia carbody with its engine air louvers mounted along the top edge of the long hood, separated roof fans, and numberboard glass secured with clips. San Francisco, September 1969. The open space below the walkways appeared only on the four pre-production units. SP retrofitted its former demonstartors with stepped handrails along the long hood.*

Louis A. Marre collection

Carbody Phases

Although most locomotive builders do not recognize subtle locomotive carbody design changes, Fairbanks-Morse made various distinctive changes to the Train Master during its four-year production span. The following distinctions among phases represent as-built differences, since several of the changes made to later production units were retrofitted to earlier units.

Phase Ia
Dates Built: April 1953-June 1953
Quantity Built: 14

The four demonstrators, TM-1 through TM-4, as well as Delaware, Lackawanna & Western's initial order for Train Masters, were built with Phase I carbodies. These units featured engine air louvers along the upper edge of the long hood, separate cooling fans, and straight handrails running the entire length of the long hood. The cast bronze rectangular builder's plate was attached to the unit with four screws. The glass numberboards were attached with clips.

Phase Ib
Dates Built: September 1953-June 1954
Quantity Built: 49

Owners of Phase Ia Train Masters were Central Railroad of New Jersey, Reading, Southern Pacific, and Virgin-

Phase Ia - *Part of the first production TM order, Delaware, Lackawanna & Western 852 lacks a row of engine air intake louvers (below the dynamic brake resistor grid opening on the long hood, just ahead of the horn) found on all subsequent units. Hoboken, N.J., January 2, 1956.*

John D. Hahn, Jr.

Phase Ib - *SP 3025 illustrates this carbody phase with its stepped handrails. The headlight arrangement and large train numberboards (they replaced the original small numberboards) are Southern Pacific options/modifications. San Francisco, April 5, 1973.*

road of New Jersey, Delaware Lackawanna & Western, Pennsylvania, Reading, Southern, Virginian, and Wabash.

Other Distinctions

Besides these differences, two TM orders were equipped with low end platforms to better match the walkway heights of contemporary EMD and Alco models. Southern Railway's five units and Wabash's six purchased-new TMs were so equipped.

The first four Train Masters built, TM-1 through TM-4 (later becoming Wabash 550, 551, and Southern Pacific 4800, 4801) featured open space below the side walkways. This was never duplicated on any other TM order.

While FM designated all Train Masters as H24-66 models, internally, the company applied a model suffix to designate specific equipment. An H24-66S was equipped with a steam generator, while an H24-66D was equipped with dynamic brakes – an H24-66DS had both options. Canadian Pacific's unique H24-66D2S model had two steam generators.

Wabash's eight Train Masters were repowered by Alco in 1964. Although the original engine compartment hood structure remained, it was rebuilt with new access doors, Alco-designed engine compartment air intakes, and new rooftop access doors. ★

ian. The only external change between the Phase Ia and Ib units is a variable handrail height along the long hood.

Phase II
Dates Built: May 1955- June 1957
Quantity Built: 64

When FM changed to an oil bath engine air filter system, the louvers along the top edge of the carbody were eliminated and replaced by a series of louvers below the dynamic brake resistor grid opening and engine air intakes. The space between the rooftop-mounted radiator fans was eliminated, and the builder's plate, while remaining rectangular in shape, was now made of stamped aluminum and attached with six screws. The four lifting pads were redesigned, and the glass numberboards were enlarged and mounted with rubber weatherstripping. Owners of Phase II Train Masters were Canadian National, Canadian Pacific, Central Railroad of New Jersey, Delaware

Phase II - *Reading 866 illustrates the appearance of post-1954 Train Master production with a revised engine air louver arrangement, larger weatherstripped numberboards, and closely-spaced radiator fans. Tamaqua, Pa., September 1, 1960.*

Phase II - *Central Railroad of New Jersey purchased six Phase II Train Masters, all built with a specially designed classification-light housing, as seen on 2408 at Jersey City, N.J., on April 25, 1967.*

Phase II - *Only 20 Train Masters were produced outside the United States. Built by FM subsidiary Canadian Locomotive Works in Kingston, Ontario, they feature squared-off fuel tanks, a simpler twin-beam headlight design, and rooftop-mounted water expansion tanks.*

Phase II - *Wabash 594 represents a Phase II Train Master with lower end platforms. Note the modified engine compartment hood – the unit was one of eight repowered by Alco in the mid-1960s. East St. Louis, Ill., October 31, 1964.*

Fairbanks-Morse H24-66

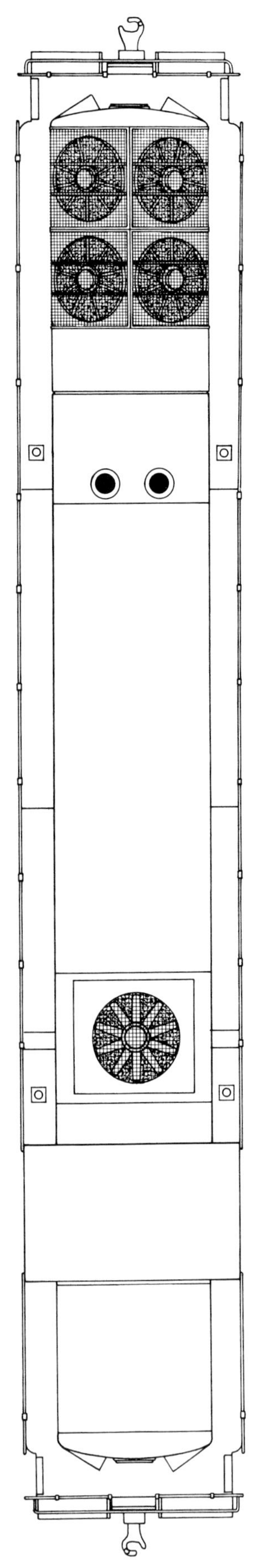

Drawing by Win Cuisinier © 1977

Notes on Fairbanks-Morse H24-66 Plans (Plan Scale 1:87)

Phase Ia - Drawn with the cab windows closed and the equipment representative of the appearance of the unit shortly after delivery.

Phase Ib - Drawn with the cab windows closed and the equipment representative of the appearance of the unit shortly after delivery.

Phase II - Drawn with the cab windows open (note that they do not retract fully into the corners of the cab). The equipment shown on this unit is representative of its final appearance).

Phase II Repowered - Drawn with the cab windows open and the equipment representative of its appearance in 1961.

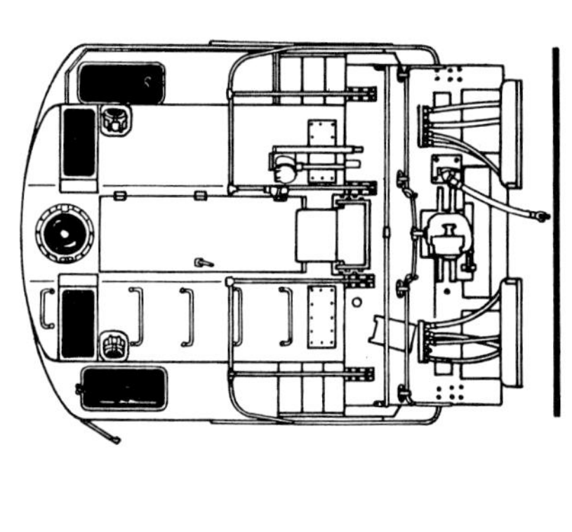

Drawings appear courtesy of Preston W. Cook and Don Dover of Extra 2200 South. The drawings were originally published in Issue 60 (April-May-June 1977) and Issue 61 (July-August-September 1977).

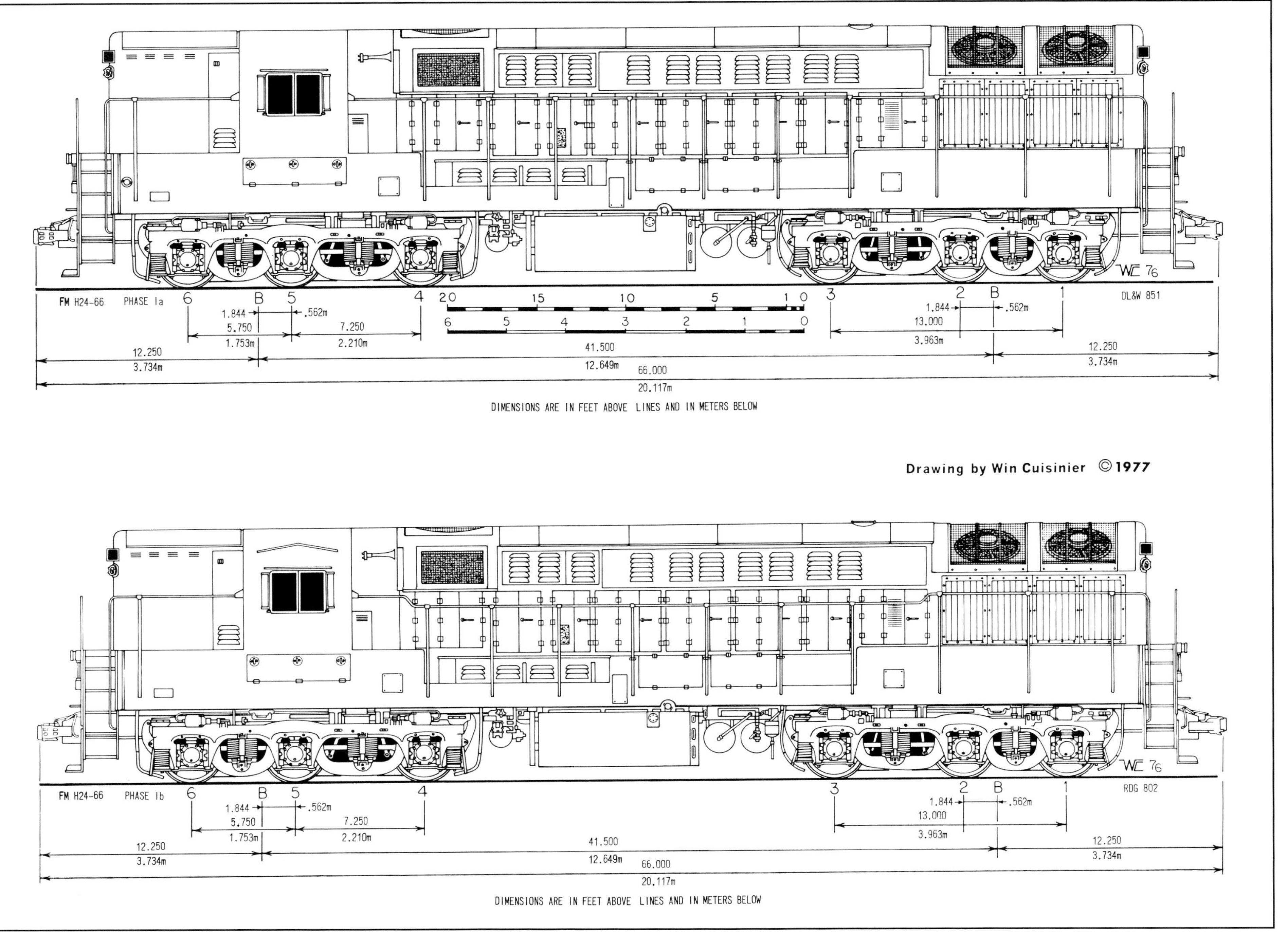

WC 76
DL&W 851
FM H24-66   PHASE Ia
DIMENSIONS ARE IN FEET ABOVE LINES AND IN METERS BELOW
Drawing by Win Cuisinier ©1977
WC 76
RDG 802
FM H24-66   PHASE Ib
DIMENSIONS ARE IN FEET ABOVE LINES AND IN METERS BELOW

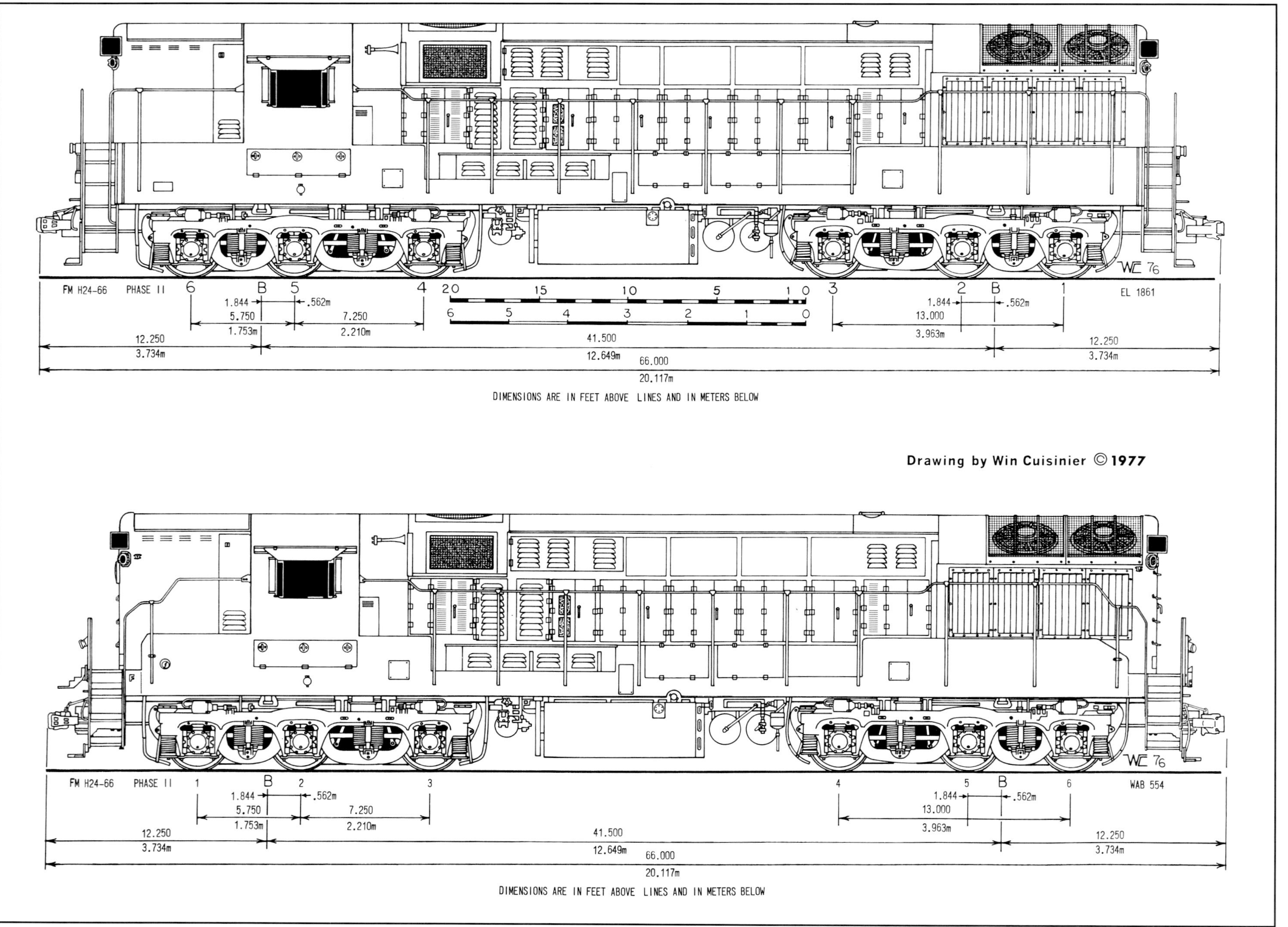
FM H24-66   PHASE II
EL 1861
WE 76
WAB 554
Drawing by Win Cuisinier ©1977
DIMENSIONS ARE IN FEET ABOVE LINES AND IN METERS BELOW
1.844   .562m
5.750   7.250
1.753m   2.210m
12.250
3.734m
41.500
12.649m   66.000
20.117m
1.844   .562m
13.000
3.963m
12.250
3.734m

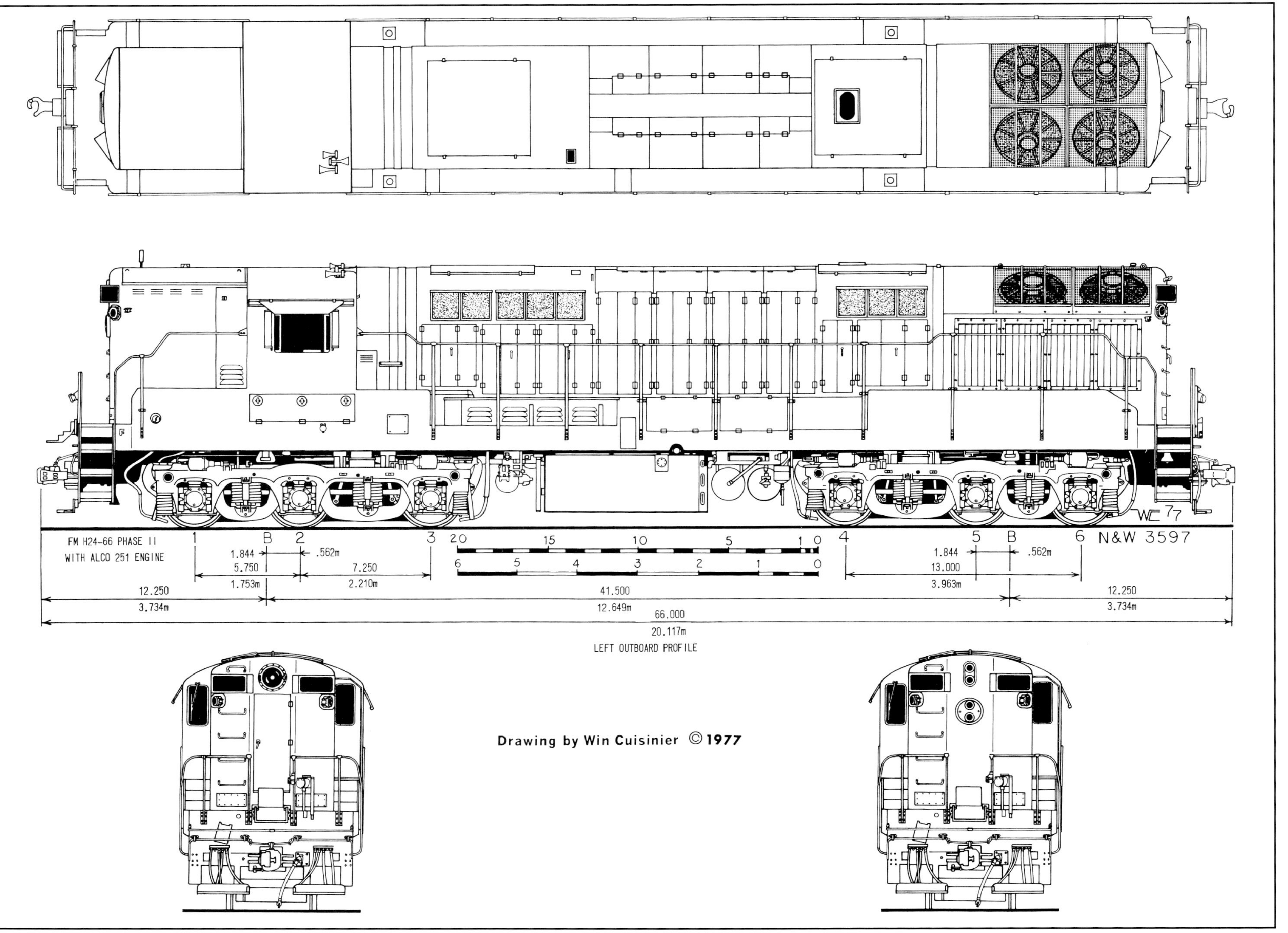

FM H24-66 PHASE II
WITH ALCO 251 ENGINE
N&W 3597
WC 77
LEFT OUTBOARD PROFILE
Drawing by Win Cuisinier © 1977

Fairbanks-Morse H24-66 Train Master Roster

RAILROAD/ OWNER	ROAD NUMBER	2nd ROAD NUMBER	3rd ROAD NUMBER	BUILDER'S NUMBER	BUILDER'S DATE	ORDER NUMBER	CARBODY PHASE	GEAR RATIO	WHEEL DIA.	OPTIONS	WEIGHT	FUEL CAPY.	WATER CAPY.	ELECT. EQUIP.	NOTES
Canadian National	3000	2900		24-L-862	7/55	C-636	II	74:18	40"	DS	375,770	1,800	2,400	GEA	
Canadian Pacific	8900			24-L-861	6/55	C-635	II	65:18	40"	DS	388,000	1,800	2,400	GEA	
Canadian Pacific	8901			2928	8/56	C-638	II	65:18	40"	D2S	389,000	1,980	2,400	GEA	
Canadian Pacific	8902			2929	8/56	C-638	II	65:18	40"	D2S	389,000	1,980	2,400	GEA	
Canadian Pacific	8903			2930	8/56	C-638	II	65:18	40"	D2S	389,000	1,980	2,400	GEA	
Canadian Pacific	8904			2931	8/56	C-638	II	65:18	40"	D2S	389,000	1,980	2,400	GEA	
Canadian Pacific	8905			2922	6/56	C-638	II	65:18	40"	DS	389,000	1,980		GEA	
Canadian Pacific	8906			2923	6/56	C-638	II	65:18	40"	DS	389,000	1,980		GEA	
Canadian Pacific	8907			2924	7/56	C-638	II	65:18	40"	DS	389,000	1,980		GEA	
Canadian Pacific	8908			2925	7/56	C-638	II	65:18	40"	DS	389,000	1,980		GEA	
Canadian Pacific	8909			2926	8/56	C-638	II	65:18	40"	DS	389,000	1,980		GEA	
Canadian Pacific	8910			2927	8/56	C-638	II	65:18	40"	DS	389,000	1,980		GEA	
Canadian Pacific	8911			2932	9/56	C-638	II	65:18	40"	DS	389,000	1,980		GEA	
Canadian Pacific	8912			2933	9/56	C-638	II	65:18	40"	DS	389,000	1,980		GEA	
Canadian Pacific	8913			2934	9/56	C-638	II	65:18	40"	DS	389,000	1,980		GEA	
Canadian Pacific	8914			2935	9/56	C-638	II	65:18	40"	DS	389,000	1,980		GEA	
Canadian Pacific	8915			2936	9/56	C-638	II	65:18	40"	DS	389,000	1,980		GEA	
Canadian Pacific	8916			2937	10/56	C-638	II	65:18	40"	DS	389,000	1,980		GEA	
Canadian Pacific	8917			2938	10/56	C-638	II	65:18	40"	DS	389,000	1,980		GEA	
Canadian Pacific	8918			2939	10/56	C-638	II	65:18	40"	DS	389,000	1,980		GEA	
Canadian Pacific	8919			2940	10/56	C-638	II	65:18	40"	DS	389,000	1,980		GEA	
Canadian Pacific	8920			2941	10/56	C-638	II	65:18	40"	DS	389,000	1,980		GEA	
Central Railroad of New Jersey	2401			24-L-849	5/54	LD-171	Ib	65:18	42"	S	386,900	1,800	2,400	GEA	
Central Railroad of New Jersey	2402			24-L-850	5/54	LD-171	Ib	65:18	42"	S	386,900	1,800	2,400	GEA	
Central Railroad of New Jersey	2403			24-L-851	5/54	LD-171	Ib	65:18	42"	S	386,900	1,800	2,400	GEA	
Central Railroad of New Jersey	2404			24-L-852	5/54	LD-171	Ib	65:18	42"	S	386,900	1,800	2,400	GEA	
Central Railroad of New Jersey	2405			24-L-853	6/54	LD-171	Ib	65:18	42"	S	386,900	1,800	2,400	GEA	
Central Railroad of New Jersey	2406			24-L-854	6/54	LD-171	Ib	65:18	42"	S	385,800	1,800	2,400	GEA	
Central Railroad of New Jersey	2407			24-L-855	6/54	LD-171	Ib	65:18	42"	S	385,800	1,800	2,400	GEA	
Central Railroad of New Jersey	2408			24-L-885	3/56	LD-191	II	65:18	42"	S	379,000	1,800	2,400	GEA	
Central Railroad of New Jersey	2409			24-L-886	3/56	LD-191	II	65:18	42"	S	379,000	1,800	2,400	GEA	
Central Railroad of New Jersey	2410			24-L-887	3/56	LD-191	II	65:18	42"	S	379,000	1,800	2,400	GEA	
Central Railroad of New Jersey	2411			24-L-888	4/56	LD-191	II	65:18	42"	S	379,000	1,800	2,400	GEA	
Central Railroad of New Jersey	2412			24-L-889	4/56	LD-191	II	65:18	42"	S	379,000	1,800	2,400	GES	
Central Railroad of New Jersey	2413			24-L-890	4/56	LD-191	II	65:18	42"	S	379,000	1,800	2,400	GEA	
Delaware, Lackawanna & Western 850		EL 1850		24-L-734	6/53	LD-147	Ia	68:15	42"	DS	379,000	1,800	2,400	WH1	
Delaware, Lackawanna & Western 851		EL 1851		24-L-735	6/53	LD-147	Ia	68:15	42"	DS	379,000	1,800	2,400	WH1	
Delaware, Lackawanna & Western 852		EL 1852		24-L-736	6/53	LD-147	Ia	68:15	42"	DS	379,000	1,800	2,400	WH1	
Delaware, Lackawanna & Western 853		EL 1853		24-L-737	6/53	LD-147	Ia	68:15	42"	DS	379,000	1,800	2,400	WH1	
Delaware, Lackawanna & Western 854		EL 1854		24-L-738	6/53	LD-147	Ia	68:15	42"	DS	379,000	1,800	2,400	WH1	
Delaware, Lackawanna & Western 855		EL 1855		24-L-739	6/53	LD-147	Ia	68:15	42"	DS	379,000	1,800	2,400	WH1	
Delaware, Lackawanna & Western 856		EL 1856		24-L-740	6/53	LD-147	Ia	68:15	42"	DS	379,000	1,800	2,400	WH1	
Delaware, Lackawanna & Western 857		EL 1857		24-L-741	6/53	LD-147	Ia	68:15	42"	DS	379,000	1,800	2,400	WH1	
Delaware, Lackawanna & Western 858		EL 1858		24-L-742	6/53	LD-147	Ia	68:15	42"	DS	379,000	1,800	2,400	WH1	
Delaware, Lackawanna & Western 859		EL 1859		24-L-743	6/53	LD-147	Ia	68:15	42"	DS	379,000	1,800	2,400	WH1	
Delaware, Lackawanna & Western 860		EL 1860		24-L-1035	11/56	LD-203	II	74:18	40"	D	376,000	1,800		GES	
Delaware, Lackawanna & Western 861		EL 1861		24-L-1036	11/56	LD-203	II	74:18	40"	D	376,000	1,800		GES	

RAILROAD/ OWNER	ROAD NUMBER	2nd ROAD NUMBER	3rd ROAD NUMBER	BUILDER'S NUMBER	BUILDER'S DATE	ORDER NUMBER	CARBODY PHASE	GEAR RATIO	WHEEL DIA.	OPTIONS	WEIGHT	FUEL CAPY.	WATER CAPY.	ELECT. EQUIP.	NOTES
Fairbanks-Morse	TM-1	WAB 550	WAB 598*	24-L-730	4/53	LD-170	Ia	63:15	42"	S	378,400	1,800	2,525	WH1	1
Fairbanks-Morse	TM-2	WAB 551	WAB 599*	24-L-731	4/53	LD-170	Ia	63:15	42"	S	378,400	1,800	2,525	WH1	1
Fairbanks-Morse	TM-3	SP 4801	SP 3020	24-L-732	5/53	LD-165	Ia	68:15	42"	DS	382,100	1,950	2,400	WH1	
Fairbanks-Morse	TM-4	SP 4800	SP 3021	24-L-733	5/53	LD-165	Ia	68:15	42"	DS	382,100	1,950	2,400	WH1	
Pennsylvania	8699	PC 6708	PC 06708	24-L-897	8/56	LD-200	II	74:18	40"	D	375,000	4,200		GES	
Pennsylvania	8700	PC 6700	PC 6799	24-L-898	8/56	LD-200	II	74:18	40"	D	375,000	4,200		GES	
Pennsylvania	8701	PC 6701	PC 06701	24-L-899	8/56	LD-200	II	74:18	40"	D	375,000	4,200		GES	
Pennsylvania	8702	PC 6702	PC 06702	24-L-900	8/56	LD-200	II	74:18	40"	D	375,000	4,200		GES	
Pennsylvania	8703	PC 6703	PC 06703	24-L-901	8/56	LD-200	II	74:18	40"	D	375,000	4,200		GES	
Pennsylvania	8704	PC 6704	PC 06704	24-L-902	9/56	LD-200	II	74:18	40"	D	375,000	4,200		GES	
Pennsylvania	8705	PC 6705	PC 06705	24-L-903	9/56	LD-200	II	74:18	40"	D	375,000	4,200		GES	
Pennsylvania	8706	PC 6706	PC 06706	24-L-904	9/56	LD-200	II	74:18	40"	D	375,000	4,200		GES	
Pennsylvania	8707	PC 6707	PC 06707	24-L-905	9/56	LD-200	II	74:18	40"	D	375,000	4,200		GES	
Reading	800			24-L-779	9/53	LD-155	Ib	68:15	42"	D	385,900	4,200		WH2	4
Reading	801	201		24-L-780	9/53	LD-155	Ib	68:15	42"	D	385,900	4,200		WH2	4
Reading	802			24-L-795	11/53	LD-158	Ib	68:15	42"	D	385,900	4,200		WH2	4
Reading	803			24-L-796	11/53	LD-158	Ib	68:15	42"	D	385,900	4,200		WH2	4
Reading	804			24-L-797	11/53	LD-158	Ib	68:15	42"	D	385,900	4,200		WH2	4
Reading	805			24-L-798	11/53	LD-158	Ib	68:15	42"	D	385,900	4,200		WH2	4
Reading	806	202		24-L-799	11/53	LD-158	Ib	68:15	42"	D	385,900	4,200		WH2	4
Reading	807			24-L-906	12/56	LD-204	II	74:18	40"	D	388,400	4,200		GES	
Reading	808	203		24-L-907	12/56	LD-204	II	74:18	40"	D	388,400	4,200		GES	
Reading	860	260		24-L-781	10/53	LD-155	Ib	68:15	42"	S	386,700	1,800	2,400	WH2	4
Reading	861			24-L-782	10/53	LD-155	Ib	68:15	42"	S	386,700	1,800	2,400	WH2	4
Reading	862	261		24-L-865	11/55	LD-185	II	68:15	42"	DS	388,400	1,800	2,400	WH3	
Reading	863	262		24-L-882	12/55	LD-185	II	68:15	42"	DS	388,400	1,800	2,400	WH3	
Reading	864	263		24-L-883	12/55	LD-185	II	68:15	42"	DS	388,400	1,800	2,400	WH3	
Reading	865	264		24-L-884	12/55	LD-185	II	68:15	42"	DS	388,400	1,800	2,400	WH3	
Reading	866			24-L-863	11/55	LD-185	II	68:15	42"	DS	386,400	1,800	2,400	WH3	
Reading	867	265		24-L-864	11/55	LD-185	II	68:15	42"	DS	386,400	1,800	2,400	WH3	
Southern (CNO&TP)	6300			24-L-856	5/66	LD-179	II	74:18	40"	DL	362,900	1,800		GEA	
Southern (CNO&TP)	6301			24-L-857	5/66	LD-179	II	74:18	40"	DL	362,900	1,800		GEA	
Southern (CNO&TP)	6302			24-L-858	6/66	LD-179	II	74:18	40"	DL	362,900	1,800		GEA	
Southern (CNO&TP)	6303			24-L-859	6/66	LD-179	II	74:18	40"	DL	362,900	1,800		GEA	
Southern (CNO&TP)	6304			24-L-860	6/66	LD-179	II	74:18	40"	DL	362,900	1,800		GEA	
Southern Pacific	4802	3022		24-L-791	12/53	LD-164A	Ib	68:15	42"	DS	379,320	1,950	2,400	WH3	5
Southern Pacific	4803	3023		24-L-792	12/53	LD-164A	Ib	68:15	42"	DS	379,320	1,800	2,400	WH3	5
Southern Pacific	4804	3024		24-L-793	12/53	LD-164A	Ib	68:15	42"	DS	379,320	1,800	2,400	WH3	5
Southern Pacific	4805	3025		24-L-794	12/53	LD-164A	Ib	68:15	42"	DS	379,320	1,800	2,400	WH3	5
Southern Pacific	4806	3026		24-L-803	12/53	LD-164A	Ib	68:15	42"	DS	379,320	1,800	2,400	WH3	5
Southern Pacific	4807	3027		24-L-804	12/53	LD-164A	Ib	68:15	42"	DS	379,320	1,950	2,400	WH3	5
Southern Pacific	4808	3028		24-L-805	12/53	LD-164A	Ib	68:15	42"	DS	379,320	1,800	2,400	WH3	5
Southern Pacific	4809	3029		24-L-806	12/53	LD-164A	Ib	68:15	42"	DS	379,320	1,950	2,400	WH3	5
Southern Pacific	4810	3030		24-L-787	11/53	LD-164B	Ib	68:15	42"	DS	379,320	1,800	2,400	WH3	5
Southern Pacific	4811	3031		24-L-788	11/53	LD-164B	Ib	68:15	42"	DS	379,320	1,800	2,400	WH3	5
Southern Pacific	4812	3032		24-L-789	11/53	LD-164B	Ib	68:15	42"	DS	379,320	1,800	2,400	WH3	5
Southern Pacific	4813	3033		24-L-790	11/53	LD-164B	Ib	68:15	42"	DS	379,320	1,800	2,400	WH3	5
Southern Pacific	4814	3034		24-L-800	2/54	LD-164B	Ib	68:15	42"	DS	379,320	1,800	2,400	WH3	5

RAILROAD/ OWNER	ROAD NUMBER	2nd ROAD NUMBER	3rd ROAD NUMBER	BUILDER'S NUMBER	BUILDER'S DATE	ORDER NUMBER	CARBODY PHASE	GEAR RATIO	WHEEL DIA.	OPTIONS	WEIGHT	FUEL CAPY.	WATER CAPY.	ELECT. EQUIP.	NOTES
Southern Pacific	4815	3035		24-L-801	2/54	LD-164B	lb	68:15	42"	DS	379,320	1,800	2,400	WH3	
Virginian	50	N&W 150		24-L-807	3/54	LD-167-1	lb	74:18	40"	D	394,500	1,780		GEA	
Virginian	51	N&W 151		24-L-808	3/54	LD-167-1	lb	74:18	40"	D	394,500	1,780		GEA	
Virginian	52	N&W 152		24-L-809	3/54	LD-167-1	lb	74:18	40"	D	394,500	1,780		GEA	
Virginian	53	N&W 153	See Note 1	24-L-810	3/54	LD-167-1	lb	74:18	40"	D	394,500	1,780		GEA	
Virginian	54	N&W 154		24-L-811	3/54	LD-167-1	lb	74:18	40"	D	394,500	1,780		GEA	
Virginian	55	N&W 155		24-L-812	3/54	LD-167-1	lb	74:18	40"	D	394,500	1,780		GEA	
Virginian	56	N&W 156		24-L-813	3/54	LD-167-1	lb	74:18	40"	D	394,500	1,780		GEA	
Virginian	57	N&W 157		24-L-814	3/54	LD-167-1	lb	74:18	40"	D	394,500	1,780		GEA	
Virginian	58	N&W 158		24-L-838	4/54	LD-167-1	lb	74:18	40"	D	394,500	1,780		GEA	
Virginian	59	N&W 159		24-L-839	4/54	LD-167-1	lb	74:18	40"	D	394,500	1,780		GEA	
Virginian	60	N&W 160		24-L-840	4/54	LD-167-1	lb	74:18	40"	D	394,500	1,780		GEA	
Virginian	61	N&W 161	See Note 1	24-L-841	4/54	LD-167-1	lb	74:18	40"	D	394,500	1,780		GEA	
Virginian	62	N&W 162		24-L-842	4/54	LD-167-1	lb	74:18	40"	D	394,500	1,780		GEA	
Virginian	63	N&W 163		24-L-843	4/54	LD-167-1	lb	74:18	40"	D	394,500	1,780		GEA	
Virginian	64	N&W 164		24-L-844	4/54	LD-167-1	lb	74:18	40"	D	394,500	1,780		GEA	
Virginian	65	N&W 165	See Note 1	24-L-845	5/54	LD-167-1	lb	74:18	40"	D	394,500	1,780		GEA	
Virginian	66	N&W 166		24-L-846	5/54	LD-167-1	lb	74:18	40"	D	394,500	1,780		GEA	
Virginian	67	N&W 167	See Note 1	24-L-847	5/54	LD-167-1	lb	74:18	40"	D	394,500	1,780		GEA	
Virginian	68	N&W 168		24-L-848	5/54	LD-167-1	lb	74:18	40"	D	394,500	1,780		GEA	
Virginian	69	N&W 169	See Note 1	24-L-1037	5/57	LD-210	II	74:18	40"	D	396,600	1,780		GES	
Virginian	70	N&W 170		24-L-1038	5/57	LD-210	II	74:18	40"	D	396,600	1,780		GES	
Virginian	71	N&W 171		24-L-1039	6/57	LD-210	II	74:18	40"	D	396,600	1,780		GES	
Virginian	72	N&W 172		24-L-1040	5/57	LD-210	II	74:18	40"	D	396,600	1,780		GES	
Virginian	73	N&W 173		24-L-1041	6/57	LD-210	II	74:18	40"	D	396,600	1,780		GES	
Virginian	74	N&W 174		24-L-1042	6/57	LD-210	II	74:18	40"	D	396,600	1,780		GES	
Wabash	552	592	See Note 2	24-L-891	4/56	LD-194	II	74:18	42"	SL	378,520	1,800	1,800	GES	
Wabash	552A	553	See Note 2	24-L-892	4/56	LD-194	II	74:18	42"	SL	378,520	1,800	1,800	GES	
Wabash	553	593	See Note 2	24-L-893	4/56	LD-194	II	74:18	42"	SL	378,520	1,800	1,800	GES	
Wabash	553A	555	See Note 2	24-L-894	4/56	LD-194	II	74:18	42"	SL	378,520	1,800	1,800	GES	
Wabash	554	594	See Note 2	24-L-895	4/56	LD-194	II	74:18	42"	SL	378,520	1,800	1,800	GES	
Wabash	554A	557	See Note 2	24-L-896	4/56	LD-194	II	74:18	42"	SL	378,520	1,800	1,800	GES	

GENERAL NOTES:
a. Options: D - Dynamic brakes
            S - Steam generator
            2S - Two steam generators
            L - Low end platforms
b. Electrical components are: WE1 - Westinghouse, 498A2 main generator, 370DEZ traction motors
            WE2 - Westinghouse, 498AZ main generator, 370DEZ traction motors
            WE3 - Westinghouse, 498BZ main generator, 370DEZ traction motors
            GEA - General Electric GT567C1 main generator, 752E traction motors, amplidyne excitation
            GES - General Electric GT567C1 main generator, 752E traction motors, static excitation

NOTES:
1. See page 84 for complete Virginian renumberings.
2. See page 88 for complete Wabash renumberings.

Larry G. Russell

By the time this photograph was taken at Montreal, Canadian National 2900 had been demoted to yard service, where mechanical breakdowns were not as critical, with maintenance personnel close at hand.

Canadian National

Number of Units: 1

Road Number: 3000

Year Built: 1955

Since the major Canadian railroads were a few years behind their American counterparts in dieselization, Fairbanks-Morse had hoped to gain a foothold for large orders by selling a single unit to each of the major carriers, Canadian Pacific and Canadian National.

With both railroads having sizable fleets of Canadian Locomotive Works-built locomotives on their rosters, it seemed natural that the carriers would be interested in the latest FM/CLC design. Consequently, Canadian National and Canadian Pacific each ordered one unit to test the merits of a 2,400-horsepower locomotive. Since production space at CLC was occupied with a CN order for 18 H16-44s and a CP order for 10 H16-44s, the Train Masters were built at Beloit. As a result, both Train Masters carried builder's plates from both Fairbanks-Morse and Canadian Locomotive Co., FM's Canadian subsidiary. The CP unit was released first, while the CN unit followed a month later, carrying a builder's date of July 1955.

CN's first road diesels were six EMD F3s acquired in 1948. Between 1950 and 1953, the carrier began adding 76 General Motors Diesel Ltd.-built (EMD's Canadian subsidiary) F7s, 26 Canadian Locomotive Works-built C-Liners, and 48 Montreal Locomotive Works-built FA/FBs to its freight fleet. Because CN was government-controlled, orders had to be split among major Canadian locomotive builders.

It appeared that Canadian National was well on its way to dieselizing all of its mainline freight operations, having developed a five-year plan that called for steam to be retired by the end of 1956. It was against this backdrop that FM, with its first Train Master nearing completion, saw an opportunity to promote its high-horsepower road switcher. Because CN was already a good FM/CLC customer – the road had purchased not only C-Liners, but also 30 H12-64 road switchers between 1951 and 1953 – and because of government dictates, FM thought it had a good shot at placing the H24-66 on the CN roster.

Delivered on August 18, 1955, number 3000 was tried in a variety of assignments from one end of the CN system to the other. Equipped with a Vapor Corp. Model 4740 steam generator and dynamic brakes, it was used in passenger

Tom Peebles/Larry G. Russell collection

Judging from the shiny paint and spotless builder's plates, 3000 was on an early run at Truro, Nova Scotia, in 1955. Although it was tried in a variety of assignments, including passenger service, CN could not seem to find a use for the TM's 2,400 horsepower.

Since Canadian National's only Train Master carried the road number 3000 for only 10 months, photographs are rare. Although the camera equipment carried by many railfan photo-graphers left much to be desired, this view at Moncton, New Brunswick, on December 11, 1955, shows the left-side details of the long hood. Both the CN and CP units were equipped with a modified end step arrangement to allow crew members to travel safely between the Train Master and other locomotive models.

CLC units after the TM was delivered, they were lower-horsepower H12-64s (20 units) and H16-44s (18 units). Unlike CP, CN could find no real use for a single-engine, 2,400-horsepower locomotive model – it found building locomotive consists with 1,600- to 1,750-horsepower units to be a more flexible approach.

By the early 1960s, CN 2900 was reassigned to Toronto for commuter service between Toronto and Hamilton, Ontario. In 1962, it was reassigned to Montreal, where it held down the Montreal-to-St. Rosalie Junction passenger run. Out of place on a roster dominated by four-axle units, the unit was equipped with hump control and assigned to the Mimico Yard hump at Toronto in April 1964.

Retired from service on February 15, 1966, the Train Master had retained its green and mustard yellow color scheme to the end. Only the lettering placement changed, with the Canadian National road name being moved from the center of each side to a location between the top hood louvers. ★

service and also put in a stint on the Port Arthur (now Thunder Bay), Ontario, ore dock. Built with a Phase II carbody, it weighed 375,770 pounds and was assigned CN class CRG-24a. Initially, it was assigned to the Manitoba District, where it was renumbered in June 1956 to 2900 to make room for a group of renumbered Alco/Montreal Locomotive Works RS-3s.

While CN went about dieselizing in the mid- to late 1950s (steam operations officially ended in April 1960), its model of choice to finish converting freight operations was GMD's GP9 model. Although CN bought additional

Although most photographs of the 2900 taken in the 1960s show the unit operating solo, it was occasionally pressed into road service mated to other locomotives, as seen in this view at Mimico Yard on June 20, 1964.

Canadian Locomotive Co.-built H24-66 8905 poses at Winnipeg, Manitoba, on August 22, 1957, in its as-built configuration with the long hood designated as front.

Canadian Pacific

Number of Units: 21

Road Numbers: 8900-8920

Years Built: 1955, 1956

Like Canadian National, Canadian Pacific owned a large fleet of Canadian Locomotive Co.-built locomotives. Its association with the Fairbanks-Morse subsidiary began on August 1, 1951, when CLC out-shopped its first streamlined cab units, painted in a Canadian Pacific-like scheme. Successful demonstration runs of these CFA16-4 C-Liners led to their purchase and to CP's acquisition of 26 more C-Liners.

When CP adopted the road-switcher carbody style on new locomotive purchases in 1954, CLC offered the H16-44 model – basically a 1,600-horsepower C-Liner surrounded by sheet metal work that shared features with FM's new

To accommodate two steam generators, 8903 was one of four Train Masters built with a full-width short hood. Only the passenger-equipped H24-66s carried CP's beaver herald on the front end. Kenora, Ontario, August 23, 1959.

Acquiring a fleet of 28 C-Liners between 1951 and 1954, Canadian Pacific then switched to the road-switcher version of the same model, the H16-44. After purchasing 10 H16-44s in 1955, CP opted for 20 H24-66s in 1956. Although not in the carrier's original dieselization plan, 10 additional H16-44s were also delivered in 1956, the result of a canceled GMD GP9R order (due to a prolonged strike at GMD in 1955-56). Orders for the H24-66 were never repeated, although 20 more H16-44s arrived in 1957. CP then chose GMD to supply the locomotives necessary to complete dieselization.

Train Master carbody. In 1954, CP placed an order with CLC for 10 of the new H16-44s and one H24-66 to test the concept of a 2,400-horsepower power plant in a single unit.

Completed in June 1955, the CP unit, numbered 8900, entered service on July 13, 1955. Built on FM/CLC contract C-635, the unit was equipped with dynamic brakes and a single Vapor Corp. Model 4740 steam generator. Its carbody used a Phase II body arrangement, and the locomotive carried General Electric electrical rotating equipment. With its long hood designated as front, the maroon-painted unit featured a gray-painted radiator section that led to a gray stripe running the length of the carbody and around the short hood. Four yellow stripes were applied to the nose of the long hood. The engine number was painted in a gray panel below the cab window, while the side sills, trucks, and frame were painted black. Yellow diagonal safety stripes were applied to both pilots.

In an era when four-axle 1,500/1,600-horsepower locomotives averaging 51 to 56 feet in length were the "standard," the 66-foot-long H24-66 must have been an impressive sight. CP considered deploying the unit on the eastern end of the system, but when it was mated with either another Train Master or even a 1,600-horsepower H16-44, the pair's tonnage capabilities, expressed in train lengths, exceeded siding capacities at the time. The unit's higher horsepower rating also precluded its use on eastern passenger trains. Tests made with 8900 and dynamometer car number 62 during July 1955 between Montreal and St. John, New Brunswick, on Trains *41/42*, the *Atlantic Limited*, showed that with the use of

Although painted with the long hood designated as front, 8900 had yet to be repainted after being reconfigured to operate with the short hood leading. Coquitlam, British Columbia, March 1963.

Now featuring a revised paint scheme, 8900 is at Trail, British Columbia, on August 9, 1965. To avoid a build-up of ice and snow on the handbrake rigging, the chain is enclosed in a protective sleeve.

one Train Master, the train was under-powered, but when a second unit was added, the train was overpowered. (In 1954, these trains were powered by CP's only 4-8-4s, K-class engines 3100 and 3101.)

Another problem was the TM's long wheelbase – the tri-mount truck was too large for the carrier's newly installed drop tables at St. Luc (Montreal) and Chapleau (Ontario), CP's primary diesel locomotive servicing facilities in the east.

right, *Although built to the same specifications as its U.S. counterpart, the CLC-built Train Masters featured a simplified twin-beam headlight housing, a brake wheel that was countersunk into the front of the short hood, a squared-off fuel tank, and rooftop water expansion tanks. Compare the details of 8900 above with 8901 and 8916 riding the Revelstoke, British Columbia, turntable on May 1, 1962.*

H24-66 8905, at Coquitlam, British Columbia, on July 20, 1963, was set aside for preservation by Canadian Pacific when it was retired on October 6, 1976.

With the cab controls reconfigured for short-hood-first operation, 8917 is at Revelstoke, British Columbia, on July 17, 1963. CP specified that three-chime horns be mounted on the cab roof instead of the two single-chime horns found on Beloit-built TMs. CP 8914, below, is at Revelstoke, British Columbia, on April 28, 1962.

During their last years of service, lettering on the TM's carbody faded to the point that it eliminated any clue as to the ownership of the locomotive. Nelson, British Columbia.

The carrier looked west for a place to use the TM's 2,400 horsepower.

Between Calgary, Alberta, and Revelstoke, British Columbia, CP tackled some of the roughest territory on its system. This 265-mile section featured maximum grades of 2.43 percent and trains typically drew sets of four 1,500- to 1,750-horsepower diesels. CP calculated that a three-unit set of Train Masters could replace these sets and handle 12-1/2 percent more tonnage.

Other assignments considered for the TM were fast freight service on CP's Prairie Region, and passenger service between Medicine Hat, Alberta, and Vancouver, British Columbia, on Trains 67/68, the *Kootenay Express*, which covered CP's southern route between these points (the streamlined transcontinental *Canadian* used the northern line through Calgary). During severe winter conditions, three units were required for their combined steam heating capabilities, although the extra horsepower of the third unit was not needed. As a result, CP's 1956 order for 30 H16-44s was changed to 20 H24-66s, four of which were equipped with two steam generators for passenger service. This resulted in a cost savings of $483,000 for the same aggregate horsepower.

Production of the 20-unit order, assigned CLC contract number C-638, began at the CLC plant in Kingston, Ontario, in early 1956, with the first unit released in June of that year. Assigned CP class DRS-24b, DRS-24c, and DRS-24d (Diesel Road Switcher, 2,400 horsepower, and the lower-case letter stood for order sequence – 8900 was assigned class DRS-24a), the units came geared for 75 mph operation. Like the demonstrator, 8901-8920 came equipped for long-hood-front operation, with the painting layout matching this arrangement. To accommodate the twin Vapor Corp. Model 4625 steam generators, 8901-8904 were built with a full-width short hood that matched the cab width.

Initial assignments for the 21-unit TM fleet included five units assigned to Calgary (Alyth shops) for Calgary-

Paterson-George collection/courtesy of Overland Models

Flying "extra'" flags, CP 8917 and 8918 make a setout at Kenora, Manitoba, during the summer of 1958.

Vancouver and Medicine Hat-Vancouver passenger service, with the balance of the H24-66s assigned to freight service in the Pacific and Prairie Regions. The units not only displaced steam locomotives such as CP's famed class T1 Selkirk 2-10-4s, but also Montreal Locomotive Works RS-3s, which were reassigned to the Eastern Region. The units remained in these assignments during the 1950s and early 1960s. For a period of time, the five steam-generator-equipped Train Masters, 8900-8904, were assigned to Winnipeg, Manitoba, and often hauled local overnight coach-and-sleeper passenger Trains 57/58 to and from Moose Jaw, Saskatchewan. At first, the TMs were mated only to other TMs and CLC-built locomotives, but in 1956, CP adopted a universal multiple-unit control system that allowed the units to be used with other makes and models, although some minor multiple-unit incompatibilities still existed.

Starting in 1959, the units were changed to short-hood-front operation. This work included removing the steam generators from 8900-8904, and on 8901-8904, the short hoods were cut back to standard width. The painting arrangement was changed to gray short hoods with three yellow stripes and gray cabs, with gray numbers on the cab sides under the windows in the maroon section. It was at this time that all but five of the H24-66s were assigned to the Prairie Region and based at Winnipeg for maintenance. They were assigned to move grain from Winnipeg to the Great Lakes port of Thunder Bay.

In the fall of 1966, CP decided to retire the 20 remaining Train Masters – one unit, 8902, was scrapped in late 1965. But no action occurred until 1968, when 14 units left the roster. Eight units were scrapped at Ogden shops (8907, 8908-8910, 8912, 8916, 8918, and 8920) and six (8906, 8911, 8913-8915, 8919) were sold to Striegel Supply & Equipment in Baltimore. Traction motors from the eight units scrapped, including 8902, were used as trade-in credit on an eight-unit order of Montreal Locomotive Works C-630s, 4500-4507. But a locomotive shortage suspended the Train Master retirement process, and the remaining units were concentrated in southern British Columbia, where they joined the balance of Canadian Pacific's CLC-built locomotive fleet, which was assigned to Nelson for maintenance.

In early 1972, 8901, 8909, and 8917 were scrapped at Angus (Calgary), while 8903 survived until April 2, 1974, when it was sold to United Railway Supply in Montreal after a brief stint in transfer service at Montreal. URS sold the engine and generator to Preco Engineering Co., in Houston, Texas, on November 6, 1974, and subsequently scrapped the balance of the unit.

Three Train Masters (8900, 8904, and 8905) were assigned to Warfield, British Columbia, for operation on a three-mile branch that serviced a metallurgical plant at Tadanac and a fertilizer plant at Warfield. The line featured a 4.18 percent grade and sharp curves. For this service, the trucks were modified to allow more play for the curves. This unique role served as a kind of reprieve, and these units were the last H24-66s retired from the CP roster. After being stored at Calgary in late 1975, they were removed from the roster on June 10, 1976, thus ending the Train Master era on Canadian Pacific. Unit 8905 later became part of the collection of the Canadian Railway Museum at Delson, Quebec, and is the sole surviving H24-66. ★

Richard O. Adams

H24-66 8904 switches a smelter at Trail, British Columbia, on September 17, 1972.

Martin S. Zak

On the property less than six months, 2411 displays the as-built appearance of CNJ's 13 H24-66s at Jersey City, N.J., on September 8, 1956.

Central Railroad of New Jersey

Number of Units: 13

Road Numbers: 2400-2413

Years Built: 1954, 1956

Central of New Jersey was an early Fairbanks-Morse customer, acquiring 14 FM road-switchers in the late 1940s for commuter service out of its terminal on the Hudson River waterfront in Jersey City, N.J. Passengers took CNJ ferryboats to and from Manhattan, connecting with trains that operated west on the railroad's main line to Phillipsburg, N.J., and beyond, or south on the New York & Long Branch Railroad to north Jersey coast resort and bedroom communities along the Atlantic shore.

CNJ had already established itself as a pioneer in the use of diesel locomotives, both in switching and commuter service. In 1925, the road had bought the nation's first commercially successful diesel-electric switcher, boxcab unit 1000, from a consortium of American Locomotive Co., General Electric, and Ingersoll-Rand. From 1946 to 1948, the road acquired six double-ended Baldwin DRX-6-4-2000 cab units, which it touted as being the first diesels in the world to be bought specifically for commuter service. Jersey Central picked up its first Fairbanks-Morse units in September 1948 when it bought a former H15-44 demonstrator unit, number 1500, and ordered 13 more production models for delivery in early 1949. Four additional FM road switchers, this time 1,600-horsepower H16-44s, came in July 1950

Diesel Era collection

Demoted to freight service, 2401 and 2405 pause at Jersey City, N.J., on September 14, 1963. Although the CNJ fleet was equipped with dual controls, the long hood was designated as the front of Jersey Central's Train Masters.

R. S. Short/The Garys collection

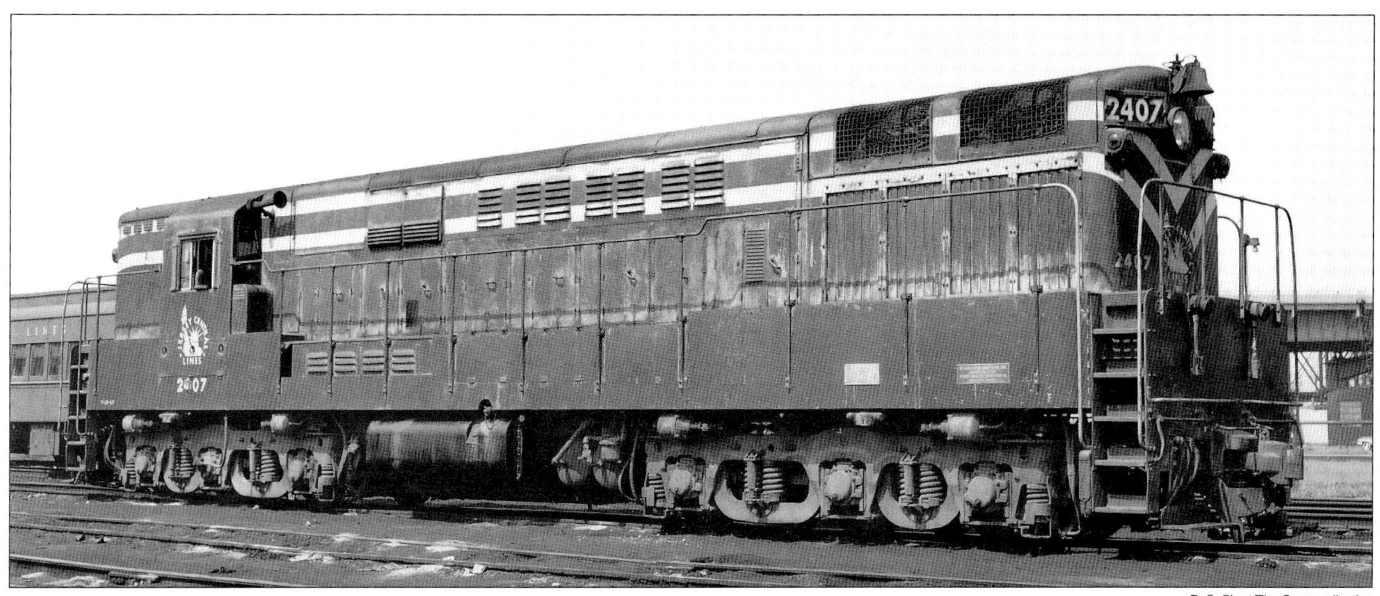

Laying over at Communipaw in Jersey City, N.J., in September 1964, 2407 displays the Phase Ib carbody that was applied to units 2401-2407. While most Train Masters were built with single-lens classification lights, CNJ specified a two-lens version that resulted in a larger, protuding housing.

to dieselize more commuter runs. With this opposed-piston power experience, it isn't surprising that Jersey Central considered H24-66s for commuter service – especially after demonstrators TM-1 and TM-2 showed off their pulling capacity and rapid acceleration on CNJ lines during their tour in 1953.

FM contract number LD-171 specified that seven H24-66 units be delivered between May and July 1954, and they were intended to kill the fires on Jersey Central's remaining steam power. By April 1954, CNJ was advertising that its passenger service was completely dieselized; on July 1 of that year, the road announced that it was 100 percent diesel-powered, and a final steam railfan trip ran on July 11 with T38-class 4-6-0 camelback engine 774. Among the passenger steamers that CNJ retired during this time were 700-series 4-6-0 Ten-Wheeler camelbacks (Baldwin, 1910-1918), which had handled many com-

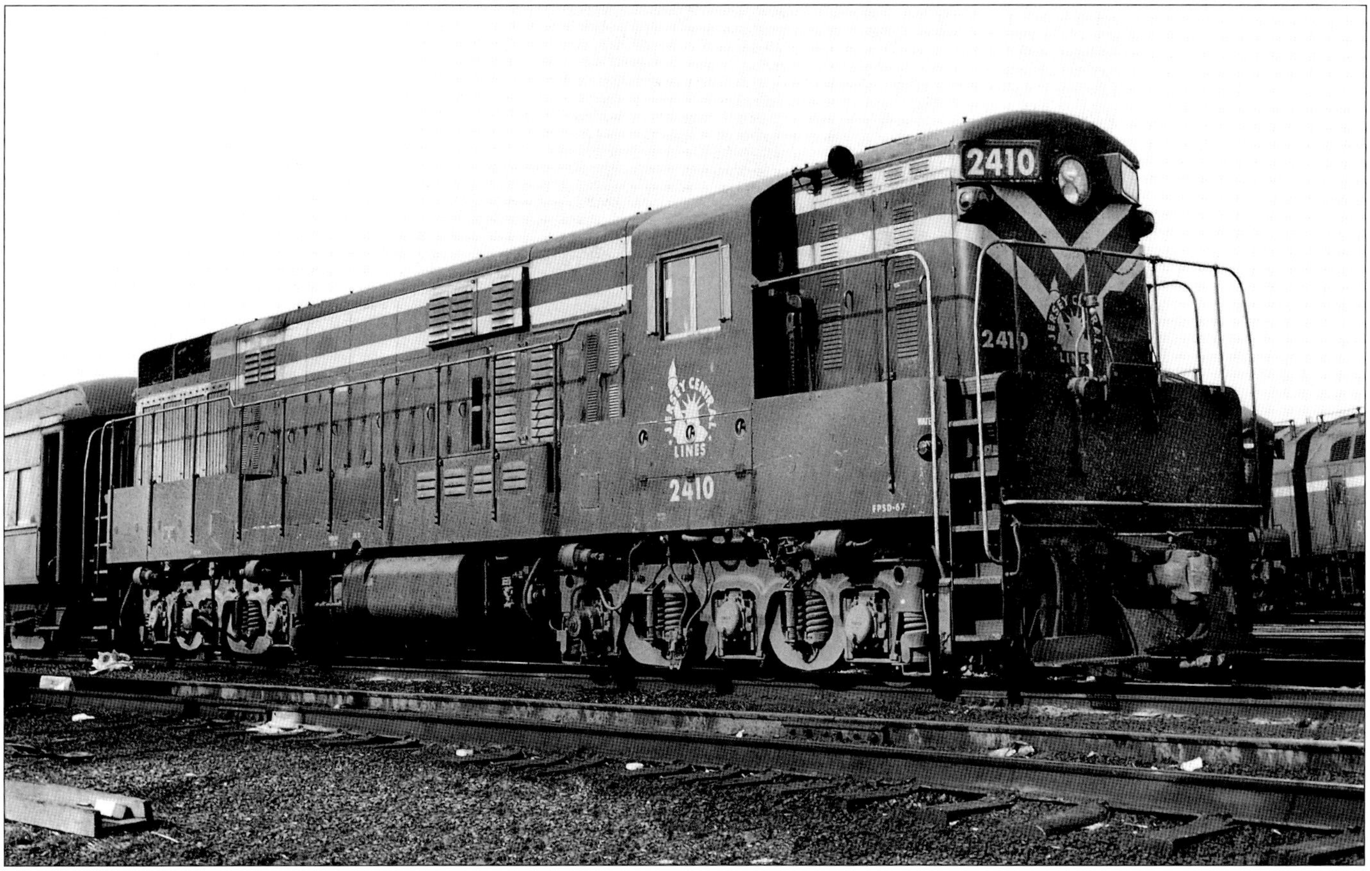

Bruce R. Meyer

One of the more common assignments was the 39.4-mile-long New York & Long Branch, a Central Railroad of New Jersey and Pennsylvania Railroad joint operation. By the early 1960s, 13 daily PRR trains each way were traversing this commuter route, the northern New Jersey line, with CNJ's nine trains each way handled by H24-66s like 2410 at Bay Head Junction, N.J., on February 10, 1963.

Double-headed Train Masters were a rarity on CNJ, especially when hauling a four-car train. Either 2413 failed enroute or is being dead headed to its assignment after a trip through Elizabethport shops. The train was southbound at Elizabethport, N.J., in November 1966.

Both Reading and Jersey Central Train Master trimount trucks featured modifications that included redesigned spring holders. Communipaw, Jersey City, N.J., October 1964.

Wearing the carrier's simplified solid green paint scheme, 2402 is at Raritan, N.J., on September 25, 1968.

muter runs, and 800-series 4-6-2 Pacifics (Baldwin, 1918-1930) that powered commuter and longer runs. While the carrier's 1,500- and 1,600-horsepower diesel units were adequate to handle six- to eight-car commuter and passenger runs, the road's longer 12-car commuter trains required pairs of such units. With the Train Master's extra horsepower, CNJ reasoned that a single TM could reduce the use of doubleheaded H15/16-44s, Alco RS-3s, and EMD GP7s.

Equipped with General Electric electrical gear, units 2401-2407 came painted in CNJ's standard olive green with yellow striping. Each locomotive carried 2,400 gallons of water for steam heat and 1,800 gallons of fuel. Built with Phase Ib carbodies, they featured dual control stands for bi-directional running, a Leslie Model A-200 horn, and a head-end lighting generator in addition to the steam generator. They were assigned CNJ class FPSD-67 – Freight, Passenger, Switching Diesel, 66,500 pounds continuous tractive effort (rounded up to the nearest thousand).

They were not equipped with dynamic brakes. Trackside, one could find the Train Masters assigned to longer commuter trains on the CNJ main line west to Raritan and Phillipsburg, N.J., or on the New York & Long Branch Railroad, a jointly owned (with Pennsylvania Railroad) line that extended 39.4 miles from Perth Amboy to Bay Head Junction, N.J., along the North Jersey seacoast. They also covered longer passenger runs to Allentown and Harrisburg, Pa., on Train *199*, the *Queen of the*

Valley; Train *194*, the *New York Clocker*; and Train *2194*, the *Harrisburger* (formerly the *Harrisburg Special*).

In 1954, shortly after delivery, Jersey Central experienced vibration problems with Train Masters that were used in high-speed Philadelphia-to-Jersey City, N.J. "clocker" passenger service, which was operated jointly with Reading Co. Reading was experiencing similar problems with its Train Masters, and as a result, ran a series of test runs called the "Jenkintown Tests," so named because of the town on the Reading line on which they were conducted.

As a result, FM and General Steel Castings Corp. designed a swing-bolster version of the tri-mount truck that was applied to Jersey Central's second Train Master order. FM engineer Robert Stacy supervised the FM factory road tests on CNJ units 2408-2409 on the 100 mph Milwaukee Road C&M Division between Sturtevant and Lake, Wis. During the tests, the redesigned truck reached speeds of 100 mph despite the units being equipped with only 80 mph gearing.

F. W. Trittenbach/Louis A. Marre collection

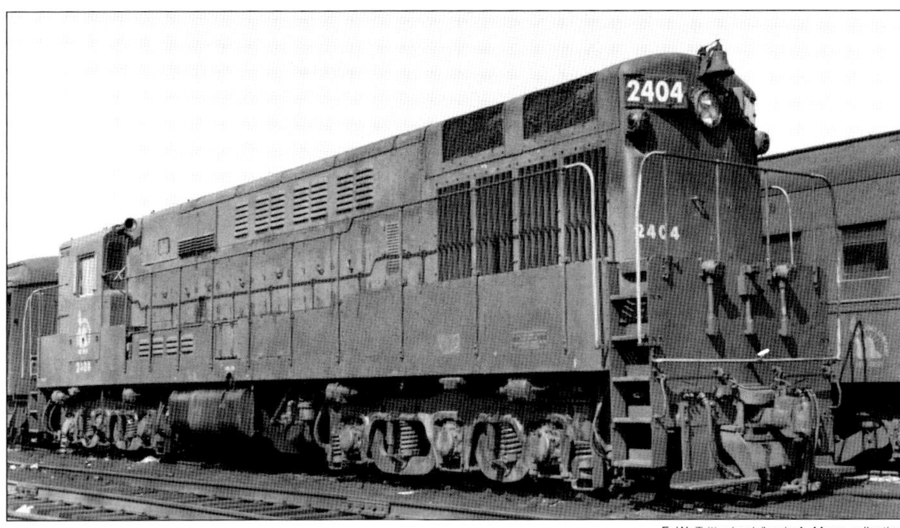

CNJ added extensions to the horn mounting brackets on some of their H24-66s to improve sound carrying qualtities. Unit 2404 is at Raritan Bay, N.J., on September 25, 1968.

The second Train Master order, delivered less than a year after the first, was purchased to replace the aging double-ended Baldwin units. Numbered 2408-2413, the Train Masters were essentially mechanical duplicates of the first order except for having Phase II carbodies. With their redesigned trucks,

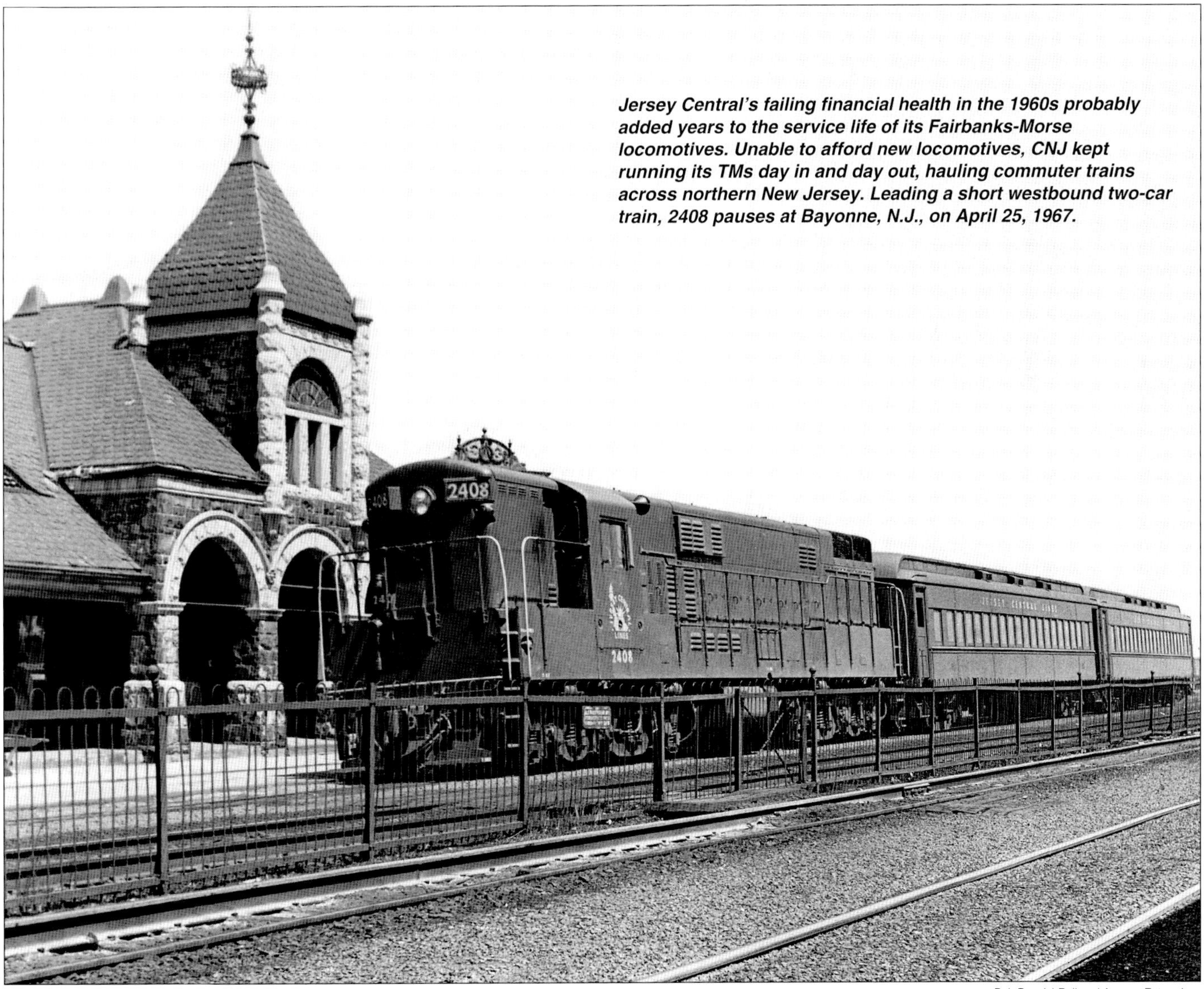

Jersey Central's failing financial health in the 1960s probably added years to the service life of its Fairbanks-Morse locomotives. Unable to afford new locomotives, CNJ kept running its TMs day in and day out, hauling commuter trains across northern New Jersey. Leading a short westbound two-car train, 2408 pauses at Bayonne, N.J., on April 25, 1967.

Bob Pennisi-Railroad Avenue Enterprises

From moving commuters to moving freight, CNJ utilized its Train Masters as intended by the builder. Leading a hopper train at Allentown, Pa., on April 23, 1966, are 2404 and 2408.

Martin S. Zak

Around 1962, Jersey Central and Baltimore & Ohio started to pool power between Jersey City and various B&O points, including Potomac Yard outside Washington, D.C., where 2403 and 2406 lead a pair of CNJ RS-3s on July 6, 1966.

Herbert H. Harwood

Mated to an SD35, H24-66 2409, at Communipaw on January 11, 1969, was working in the freight pool after being displaced from commuter service with the arrival of GP40Ps.

J. R. Quinn/Louis A. Marre collection

yard at Allentown, Pa. Operating in Jersey City-Philadelphia passenger service, CNJ Train Masters often showed up under the wide train shed at Reading Terminal in Philadelphia.

They were often assigned to nameless trains on this route, only occasionally drawing assignments to banner runs such as the *Wall Street* and *Crusader*, which usually were powered by Reading's EMD FP7s, with their higher-speed (89 mph) gearing. Jersey Central's H24-66s were mated with a variety of makes and models – they were coupled to Reading FP7s in Jersey City-Philadelphia passenger service, as well as to CNJ Alco RS-3s in freight service.

In December 1960, Jersey Central deactivated the steam generators on two Train Masters, 2400 and 2403, and they were then more-or-less permanently assigned to the freight pool. In 1963 and 1964, two additional units, 2402 and 2405, were so modified. In 1962, the carrier adopted a painting austerity program that eliminated the striping on repainted units. Many CNJ Train Masters wore this simplified solid green color scheme, which was relieved only by heralds applied to the cab sides.

By the late 1960s, commuter service had become a severe financial drain on Jersey Central. The state of New Jersey, along with CNJ's parent company, Chesapeake & Ohio (through its controlling interests in Baltimore & Ohio and Reading Co.) financed 13 new commuter locomotives, 3,000-horsepower EMD GP40P models. Delivery of these units began in December 1968, and by May 1969, all 13 Train Masters were sold for scrap to Naporano Iron & Metal Co. of Newark, N.J. ★

they were allowed a maximum speed of 70 mph, while the seven earlier units were restricted to 60 mph. They were mixed freely with the earlier H24-66s, and handled both passenger and freight duties.

On weekends, the Train Masters often ran in pairs on freight runs to the Baltimore & Ohio interchange at Philadelphia or to CNJ's major classification

Frank DiFalco/George Melvin collection

above, *While Jersey Central's RS-3s and GP7s had to be specially modified to accommodate either train lighting or cab signal equipment, the Train Master's roomy carbody allowed the head-end lighting equipment to be carried internally; the cab signal gear was mounted in the space below the left side walkway (the box with three straps just above the fuel tank filler spout). Jersey City, N.J., September 26, 1966.*

middle, *CNJ 2409 is at Raritan Bay, N.J., on October 29, 1968.*

above, *Although the TM retained the roof contour introduced on earlier FM road switcher models, the headlight sheet metal work was eliminated. H24-66 2411 and an unidentified H15-44 share trackage at Jersey City, N.J., on September 26, 1966.*

Diesel Era collection

Frank DiFalco/George Melvin collection

The first production Train Master, Lackawanna 850, poses for its builder's portrait at Beloit, Wis., in June 1953.

Delaware, Lackawanna & Western

Number of Units: 12

Road Numbers: 850-861

Years Built: 1953, 1956

This front view (the Number 1 end) of Lackawanna 850 shows the first style of numberboards used on the Train Master.

As the Delaware, Lackawanna & Western neared complete dieselization, it ordered specific units to replace steam engines in specific assignments. The carrier's 1952 order included six H16-44 road switchers, the first Fairbanks-Morse locomotives for DL&W. Even before they arrived in December, Lackawanna, a month earlier, ordered 10 Train Masters, with the intent of completing dieselization. All this took place even before the TM's engineering drawings were finished. The Train Master's impressive specifications proved to be the right unit at the right time for DL&W.

FM later prepared *The Lackawanna Story*, a booklet describing why DL&W selected the Train Master, as a marketing tool to stimulate interest in the model. It explained the decision this way:

"The Delaware, Lackawanna & Western, the first railroad to buy Train Masters, had made a study to determine motive power require-

ments for the final stage of their dieselization. The study showed they needed four conventional road passenger units in suburban service and 10 conventional 1,500- to 1,600-horsepower units of other types for road freight duty and some suburban work.

"However, the potentialities of the newly introduced Train Master caused the alert Lackawanna management to stop and take a second look. As a result, instead of 14 units as originally contemplated, the Lackawanna purchased 10 Train Masters to fill out these assignments.

"This move was certainly contrary to general practice. A railroad owning over 200 diesel units doesn't normally wind up buying 10 units of a different type as its final diesel locomotive purchase, but the Lackawanna did – and not only saved 9 percent of the initial purchase cost, but eliminated four units from

This broadside view of 850 illustrates the typical Phase Ia carbody with split roof-mounted radiator fans, straight handrails, and engine air louvers running along the top of the sides of the long hood.

its maintenance and servicing costs."

In June 1953, under FM contract LD-147, Lackawanna began to take delivery of its first Train Masters, the first production units of this model. Equipped with steam generators, the units carried passenger-series road numbers 850-859. They were painted in Lackawanna's gray and maroon passenger scheme, with a bell-shaped yellow panel on the front nose and an arrow on the edge of the running boards. They became the first, and as it turned out, the only Lackawanna road switchers to be painted in passenger colors.

Delivery of the first 10 Train Masters started on June 5, 1953, when locomotives 850 and 851 left the Beloit plant at 6:30 p.m. and were turned over to Milwaukee Road for the journey to Chicago. They were delivered to the Erie Railroad at Hammond, Ind., the following day, where they were placed in service. Both units departed Hammond on June 6 at 11:30 p.m., powering Train *98*, which consisted of 97 loads and no empties, for a total of 4,078 tons. Unit 850 arrived at Binghamton, N.Y., on June 7, was interchanged to DL&W at 9:46 a.m., and continued on to Scranton, Pa., arriving at 3:15 p.m. the same day. Its mate, 851, did not arrive at Binghamton until 9:30 a.m. on June 8, having been involved in an accident on the Erie at Hornell, N.Y.

On June 11, units 852 and 853 left Beloit, followed by 854 and 855 on June 18, and 856 and 857 on June 26. En route, both 856 and 857 derailed in Blue Island, Ill., at 5:45 a.m. on June 27, 1953, but were quickly rerailed by 10:45 a.m. Indiana Harbor Belt made minor repairs to the units before turning them over to Erie at Hammond.

The final two units, 858 and 859, departed Beloit on July 1, 1953, via Chicago & North Western. After arriving at Hammond, they were serviced and worked east on Train *98*. Departing at 11:30 p.m. on July 2, they handled 99 loads and eight empties, for a total of 5,531 tons.

For steam passenger power, the end came very quickly – in fact, it was retired the day before the first Train Master actually reached Lackawanna rails. In its 1953 annual report, DL&W stated: "Effective June 6, full dieselization was accomplished on the Boonton Branch [a line on which commuter-

Although they were used in commuter service, the Train Master's 2,400-horsepower rating was better suited for freight duties, where a pair of the units replaced three-unit sets of 1,350- to 1,500-horsepower cab units. In this view at Scranton, Pa., in 1958, 858 and 853 show the effects of mountain railroading – a heavy coating of sand dust and grime.

Ken Douglas/Louis A. Marre collection

Under the wires at Dover, N.J., on December 29, 1957, 854 demonstrates the versatility of the Train Master. From commuter service to freight and helper assignments, Lackawanna found the TM the most economical way to complete dieselization.

train frequencies had been reduced earlier in the year] and fares were raised to the higher level of the electric service on the Morristown Line and its branches to Montclair and Gladstone." Systemwide, the report noted, "since July 13 no steam engines have been operated in either yard or road service." The Boonton line, with its associated Sussex Branch to Branchville, N.J., and a portion of DL&W's original main line to Washington, N.J., was the only Lackawanna commuter route not to have been electrified in the carrier's 70-route-mile modernization of 1930-31. In commuter

service, the Train Masters replaced 1100-series Alco 4-6-2 Pacifics built in the late 'teens and early 1920s.

Fairbanks-Morse's specification sheet indicated that all 10 of the initial units were equipped with Vapor Corp. Model 4740 steam generators, Union Switch & Signal cab signal equipment, Leslie A-200-LPA aluminum horns, Barco speedometers, and Gould batteries and deadman pedals. The 42-inch wheelsets were equipped with Hyatt roller bearings and Westinghouse Model 370 traction motors. To enable them to work in multiple-unit configu-

ration with EMD models, the TMs were equipped with field-loop dynamic brakes. As the largest DL&W diesel locomotives, the Train Masters weighed 379,000 pounds fully provisioned.

Because the end platforms on the units were high – seven feet, seven inches above the rail – DL&W issued instructions shortly after delivery that crew members not attempt to cross between these units and models from EMD or Alco, which had lower end platforms.

The Lackawanna Story, written in the fall of 1954, after Lackawanna's TMs had accumulated more than a year of road service, explained that the units exceed the carrier's expectations:

"The Lackawanna has since proved that the flexibility of the 10-unit Train Master pool is never less than, and often exceeds, that which could be expected of the 14 units of several types originally considered. Since the Train Masters are not restricted to one class of service because of design or capacity limitation, purchase of these high horse-power, high utilization locomotives was the most economical and efficient way to finish the Lackawanna dieselization. It was the promise of this result that guided the selection of the Train Master.

"The Lackawanna uses a two-unit 4,800-horsepower Train Master combination for powering its symbol freights between Scranton and Hoboken - the toughest part of the Lackawanna system. In this service, the two-unit Train Masters

Martin S. Zak

The Train Masters were not only the heaviest diesel locomotives on the Lackawanna, but the only C-C truck units purchased by the carrier. They held the weight title on the post-merger Erie Lackawanna roster until 1967, when 20 SD45s arrived. Note the extended-style lettering on 859, similar to that used on FTs and other DL&W diesels, but larger than the as-delivered style. Scranton, Pa., 1958.

When Lackawanna reordered Train Masters, subtle changes had been made to the model's exterior design. Built with a Phase II carbody, the units featured larger rubber-gasketed numberboards, the headlight was mounted closer to the roofline, and a slightly different MU receptacle arrangement was specified. Internally, the units were equipped with GE-built rotating electrical equipment and oil-bath air filters.

handle the same work load as any other three-unit freight locomotive. Their 30 percent greater transmission capacity provides an extra reserve when needed for handling heavier trains – yet with only two units to service and maintain instead of three.

"From the Pocono Summit, their dynamic brake – most powerful on the Lackawanna by 33 percent – speeds controlled descents, helps move traffic faster in both directions. Some of the suburban Train Masters augment this service to help move the heavy weekend freight traffic flowing into the busy New York port area.

"In a typical month 10 Train Master units ran 61,000 unit-miles, 60 percent of which was accounted for in this heavy mountain freight service.

"The Lackawanna Train Masters work out of Hoboken an average 187 miles a day in this highly congested area. The Lackawanna has taken advantage of the two-way operating visibility of these hood-type locomotives and operates the units in both directions without turning them around.

"All the Train Masters are maintained at Scranton, 134 miles away, but they are not run back light for monthly inspections. Instead, sub- urban units are exchanged with their identical twins in passenger or freight service, and work their way west to Scranton as head end power on a passenger train, or as half of a two-unit Train Master road freight locomotive.

"The Train Master group regularly furnishes power for the *Scrantonian* and *Merchants [Express]* between Scranton and Hoboken, as well as local passenger runs on branch lines of the Morris and Essex Division. The Train Masters handle the heavier passenger pool duties in stride, and though geared for a maximum speed of 65 mph have been called out on one occasion for head-end power on the Lackawanna's famed name train, the *Phoebe Snow*."

Still, Lackawanna found a few flaws. The railroad discovered that its Train Masters were slow in accelerating, and at high speeds they seemed to lack full rated horsepower. Jointly with Fairbanks-Morse, DL&W conducted tests in November 1953, using EMD-built E8 810 and Train Master 857 on the following passenger runs: Train *26*, the *Merchants Express*, (Scranton to Hoboken), Train *11*, the *Scrantonian* (Hoboken to Scranton), Train *1301*, the *Interstate Express* (Scranton to Binghamton), Train *1915*, the *Owl* (Binghamton to Syracuse), Train *1910* , the *New York Mail* (Syracuse to Binghamton), and Train *1306*, the *Interstate Express* (Binghamton to Scranton). It was observed throughout the 12 tests that the Train Master maintained a higher rate of acceleration, but was at a disadvantage to the E8 because of its maximum allowable speed of 65 mph, while the E8 often obtained a speed of 75 mph.

Three years after the delivery of the

This rear view (the Number 2 end) of Lackawanna 860 shows the enlarged rubber-gasketed glass numberboards and simplified pilot design.

Wearing the post-merger black and yellow freight scheme, former DL&W 850 occupies trackage in Scranton, Pa., in July 1964 that is now part of the Steamtown National Historical Site. Note that the single-chime horn has been relocated to the short hood roof.

The Garys

Typically, both Lackawanna and Erie Lackawanna operated the Train Masters in pure sets, but they were equipped with multiple-unit controls that were compatible with other makes and models. Meadville, Pa., August 16, 1965.

The Garys

Wearing its third paint scheme, EL 1853 was in its final year of service at Youngstown, Ohio, on August 16, 1967. Note the addition of spark arrestors to the exhaust stacks.

The Garys

initial order, DL&W placed a follow-on order for two more H24-66s. These differed from the first order in that they were built with General Electric, instead of Westinghouse, electrical equipment. Numbered after the first order, 860 and 861 were purchased for freight service and lacked steam generators. Built in November 1956 under FM contract LD-203, these H24-66s featured Phase II carbodies and, as with the previous order, the long hood was designated as front. The delivery of these units allowed Lackawanna to assemble additional three-unit sets of Train Masters for freight service.

These last two Train Masters nearly completed Lackawanna's diesel roster,

All photos this page, The Garys

As on other railroads, first-generation minority-make diesels on EL such as the Train Masters from the first order, operated until the expiration of their 15-year equipment trusts and then were quickly disposed of. Although 1860 and 1861 were three years newer, they lasted only two more years before being retired. EL 1857 is at Cleveland on August 16, 1967.

The broadside view of EL 1857 at Meadville, Pa., on May 21, 1967, shows the Train Master's twin round exhaust stacks and a firecracker-style radio antenna mounted on the short hood peak. Although the steam generator had been deactivated by this late date, the water fill spout is visible just in front of the "Radio Equipped" stencil – possibly being used for increased fuel capacity.

The Train Master's 2,400-horsepower rating dimmed in comparison with the second-generation high-horsepower locomotives introduced in the mid-1960s from the three remaining locomotive builders, EMD, GE, and Alco. With their minority position on the roster, EL assigned the TMs to less time-sensitive roles, such as moving iron ore from the Cleveland docks to the steel furnaces around Youngstown, Ohio, and in western Pennsylvania. EL 1857 and 1853 are at Cleveland on April 16, 1967.

with the road's last units, eight SW1200s, arriving in mid-1957. These were the final additions to DL&W's roster before the carrier merged with Erie on October 17, 1960.

With the Erie-Lackawanna merger, the Train Masters were renumbered by simply adding the digit "1" to their numbers, turning 850 into 1850, and so forth. But about the time the merger took place, a nationwide traffic slump hit, and most of the TMs were placed into storage at Scranton, site of the former DL&W's major locomotive shop. The combined ledgers of DL&W and Erie for 1960 showed a deficit of $19.9 million, and in 1961, the first full year of the merger, the carrier lost $26.5 million. As the nation's economy began to rebound, the Train Masters, one by one, were returned to service at Scranton. Shortly after the merger, and before the gray and maroon Lackawanna passenger scheme was adopted as the standard for all power, EL painted all 12 Train Masters into an Erie-inspired black and yellow freight scheme. In the mid-1960s, this scheme was replaced by a gray and maroon scheme with solid yellow ends. Three of the TMs wore a wide maroon stripe – 1850, 1854, and 1861 - which extended halfway up the cab window. At least one Train Master, 1856, was never repainted into gray and maroon.

The discontinuance of Lackawanna and Erie long-haul passenger trains, and the consolidation of passenger service after the merger, left EL with a surplus of dedicated passenger locomotives in its EMD E8 and Alco PA fleets. They became sufficient to cover the remaining intercity assignments, and they also worked side-by-side with steam-equipped Alco RS-2s and RS-3s and EMD GP7s to pull nonelectrified commuter runs out of Hoboken Terminal. Thus, the Train Masters were no longer needed for commuter service, and EL reassigned them to several locations on the former Erie Railroad. At first, they operated in three-unit sets out of Hornell, N.Y. They later moved iron ore trains between Cleveland and Youngstown, Ohio.

At the age of 15, the first 10 TMs were retired and sold to Striegel Supply & Equipment Corp., and in June 1968, they were shipped to that company's facility at Baltimore. Two of these units, 1852 and 1859, were subsequently leased on September 24, 1969, to Chihuahua Pacific in Mexico, which later purchased the units in 1971.

The two remaining Train Masters, 1860 and 1861, operated out of Brier Hill Yard in Youngstown, Ohio. Retired in August 1969 and October 1970, respectively, they, too, were sold to Striegel. Their 12-cylinder opposed-piston diesel engines found a home on the used equipment market, and the rest of the units' carbodies and components were cut up for scrap. ★

One of three units repainted with a wide maroon band, 1861 idles with 1854 at Cleveland on May 15, 1966.

H24-66 1861's days were numbered when it was photographed at Youngstown, Ohio, on June 18, 1970, assigned to switching service. The last surviving Train Master of the Lackawanna fleet, 1861 would be stricken from the roster in October, ending an 18-year era of opposed-piston engines moving freight and passengers on DL&W and EL.

By March 10, 1974, former EL 1852 had been reduced to a rust hulk, stripped by Chihuahua Pacific as a parts source for its purchased-new FM road-switcher fleet. While EL 1859 was to become CH-P 535, it was never actually renumbered and probably never operated on the carrier. Both units were scrapped by the early 1980s.

Both photos, Overland Models collection

Fairbanks-Morse's first diesel locomotive customer, Milwaukee Road, also became the first railroad to serve as host for the Train Master demonstrators. Laying over at the carrier's Bensenville, Ill., engine servicing facility, TM-2 and TM-1 display their original solid orange paint scheme in April 1953. below, TM-1 and TM-2 prepare to depart Milwaukee with a passenger train.

Fairbanks-Morse

Number of Units: 4

Road Numbers: TM-1, TM-2, TM-3, TM-4

Year Built: 1953

To promote its newest engineering effort, Fairbanks-Morse Co. assembled a quartet of Train Masters for demonstration purposes. After nearly a year of advertising the TM's potential in all types of service, FM took the next step by making available to the nation's railroad decision-makers a locomotive they could get their hands on, not just read about. As the builder had done in 1947 when it introduced its H15-44, H20-44, and Erie-built models at the Railway Supply Manufacturers' Association convention in Atlantic City, N.J., FM again set the target date to introduce its new model to coincide with the Atlantic City event. But prior to the June 1953 trade show,

TM-2 and TM-1 are testing on Western Maryland in this July 8, 1953, photograph at Port Covington (Baltimore), Md.

Russell L. Wilcox

Louis A. Marre collection

TM-1 and TM-2 cross the Illinois Central with a 70-car Wabash freight on November 30, 1953, at Wabic Tower in Decatur, Ill. The pair eventually was purchased by Wabash, which found the model to be suitable for both fast freight and passenger service.

FM performed a variety of test runs on its local railroad, Milwaukee Road.

The first of these occurred in February 1953 when TM-1 began a series of shakedown runs (see photo on page 49) on the Milwaukee. At the conclusion, the pair was returned to Beloit for final adjustments and repainting. By May, TM-1 and TM-2 were ready to be formally unveiled to the public and at a ceremony at Chicago's Union Station on May 18, 1953, FM proudly announced that it had built the "world's most powerful single-engine Diesel railroad powerhouse." While TM-1 and TM-2 departed on an eastern tour, FM in late May released a second pair of Train Masters for demonstrations west of the Mississippi.

At the conclusion of their demonstrations, all four TMs were sold. TM-1 and TM-2 were purchased by Wabash in February 1954, while TM-3 and TM-4 were purchased by Southern Pacific in November 1953. Before TM-1 and TM-2 were delivered to Wabash, their dynamic brakes were removed, and they were repainted and renumbered. SP shop personnel repainted its pair of TMs and added a steam generator to TM-4. ★

1953 Train Master Demonstrations

TM-1 and TM-2

Baltimore & Ohio	Late May-Early June
Central Railroad of New Jersey	Early September
Detroit, Toledo & Ironton	Late May
Long Island	August 25-September 3
New Haven	August
New York Central	Summer
Pennsylvania	Mid-May, Early July
Reading	July 1-July 7
Virginian	Summer
Wabash	November
Western Maryland	Early June

TM-3 and TM-4

Chicago & North Western	Summer
Denver & Rio Grande Western	Summer
Duluth, Missabe & Iron Range	June 29-July 15
Illinois Central	October
Great Northern	Summer
Rock Island	Summer
Santa Fe	Summer
Southern Pacific	August-November
Union Pacific	Summer

In addition, Milwaukee Road served as a "test track" for all four demonstrators.

Wearing 800-series numbers (their builder's numbers) to avoid dispatching conflicts with Milwaukee Road locomotives during shakedown runs, Virginian H24-66s 50 and 51 are mated to Milwaukee C-Liners 24C, 24B, and 24A at Savannah, Ill., on March 23, 1954.

A Fairbanks-Morse Veteran Recalls the Train Master

When the Train Master concept was born in 1951, Bob Stacy was a young electrical engineer on the staff of Fairbanks-Morse who watched as the demonstrator units were assembled in FM's plant at Beloit, Wis. After they were completed in early 1953, he oversaw their testing, and periodically was called on to handle troubleshooting questions with customers. His 17-year career with FM lasted from 1947 to 1964.

A native of Indianapolis who had worked on the Indiana Railroad, an interurban line, from 1936 to 1941, Stacy earned the last degree in railway electrical engineering ever granted by the University of Illinois, in 1942. After Army service in World War II, he was discharged in April 1946 and joined the staff of Baldwin Locomotive Works. There, he worked on such projects as the Pennsylvania Railroad T1-class 4-4-4-4 streamlined steam locomotives, 2,000-horsepower Jersey Central DRX-6-4-2000 "babyface" double-ended passenger units, and 6,000-horsepower DR-12-8-3000 Centipedes for Seaboard Air Line.

While Baldwin's business was good, Stacy noted, "you could see their methods and philosophy [were] not fitting in with the diesel age," so after a year and a half, he quit in October 1947 to go to work for Fairbanks-Morse in Chicago.

"The diesel business was mushrooming, Fairbanks' business in locomotives was mushrooming, and I was first assigned to a group calculating performance," he said. "They knew I had a background in engineering, and locomotive testing at Baldwin – in fact, I was down on the Seaboard on those Centipedes. I was soon transferred to a semi-secret group that was designing the C-Line locomotive. In '49, we were moved to Beloit, and the C-Line locomotive came out in the very end of '49. I was in locomotive engineering, and I got involved in locomotive testing at the plant and [would also] get out on the road occasionally."

Stacy was handling C-Line service problems and modifications – his title was service-test liaison engineer – when the Train Master came on the scene: "We had a comparatively small department, and everybody worked on everything. Jack Stotz (FM's chief engineer for locomotive engineering) was a former Westinghouse man and the boss at the time. One time he called me over and told me about the project, this was in late '51. [It] was unusual, because that was almost a one-man design, by George Derrig – he was an alumnus of Buda (Co., a maintenance-of-way equipment supplier) and EMD. I still marvel that he

Painted solid orange, demonstrator TM-1 is pictured on its first road test in February 1953. The unit is just east of Beloit on Milwaukee Road trackage.

Lettering has yet to be applied to Delaware, Lackawanna & Western 856 as it, and unpainted 857, test on the Milwaukee Road east of Beloit, Wis., in June 1953.

designed that Train Master almost single-handedly.

"I watched it being built, and hung around pretty close when it was being stationary-tested. You'd check out all the electrical systems, and then the dramatic moment always comes when you start the engine up. The engine had gone through its own tests, so it had been run before. You start the engine up, and start checking out rotation of fans, and then another dramatic moment comes when you first move the locomotive. And you check rotation there, to make sure all the traction motors go forward when you put the reverser in forward, things like that. They got to be pretty good; we didn't have much [of a problem although] some of the antics at Baldwin got to be pretty wild – the locomotive going backward instead of forward, stuff like that. Then after you pass all that, you took the locomotive out light.

"We had road testing right out of Beloit. [FM] had an agreement with the Milwaukee Railroad to use their main-line tracks for light engine testing, which we did on every production locomotive.

Every road engine out went out on the road testing. That superseded the two-hour track test that was done on all switch engines to check the bearings out, the journals." Stacy tested both demonstrator sets, units TM-1/TM-2, and TM-3/TM-4, before they left Beloit on their promotional tours.

"The fact [was] that we could do it so cheaply," he said. "All we did was pay for the crew. [Testing took place] southeast of the plant, on the main line of the Racine (Wis.)-to-Savannah (Ill.) line. They called it the RSW from one of its old, old names, the Racine & South Western. It went from Racine (Wis., on Lake Michigan) to Sturtevant (on the Chicago-Milwaukee main line), through Beloit and Freeport (Ill.), and joined the (Chicago-Omaha) main line at Lanark (Ill.).

"And when [we got] a new model out, we'd extend that to include pulling some trains. On the Milwaukee, they loved us, because they were always short of power. When the gravel business came along, they didn't have any power, so they used us. There was a big pit at Beloit, so we did a lot of gravel train hauling. It was mostly, if not all, company gravel, for

ballast." These trains typically operated 54 miles southwest over the RSW line from Beloit to Lanark, then 18 miles west as far west as Savannah, Ill. Although the RSW was a secondary branch, it was maintained to 50 mph standards, Stacy said.

Among the first journeys off the Milwaukee Road with the TM-1 and TM-2 demonstrators was a trip east from Chicago over the Pennsylvania Railroad: "One run that we did was Train *44*, which was the last train [of the day] out of Union Station. It left after Train *54*, which was the *Gotham Limited* and that left at 11-something (p.m.); this left even later, 11:50. It was primarily mail and express and had maybe just one rider car, and I was in that. And of course it did a great job handling a train like that – went down to Pittsburgh, got there in the morning and I went over [to the] nearby Fort Pitt Hotel, and bunked.

"Got rested up and boarded the [return trip], must have left there about 6 o'clock at night, again, Train *45*, all mail and express, back to Chicago. One event, we broke a pin on the axle generator, which governed the transition, on one of the locomotives, it was an outfit that FM had fabricated. General Electric had one, the same in principle, but actually much heavier equipment. We'd have been better off with it. We had to manually do the transition on that locomotive all the way from Crestline (Ohio) to Chicago. We got in, otherwise uneventfully, [went] back to the diesel pit in Chicago, and then I went back to Beloit to rest up. The fact that it was an otherwise uneventful trip was what we wanted." Eventually, Pennsy bought eight H24-66s, but Stacy noted that PRR operating-department officials "put their Train Masters into transfer and yard service, [and] there wasn't anything exceptional about that. Fel-

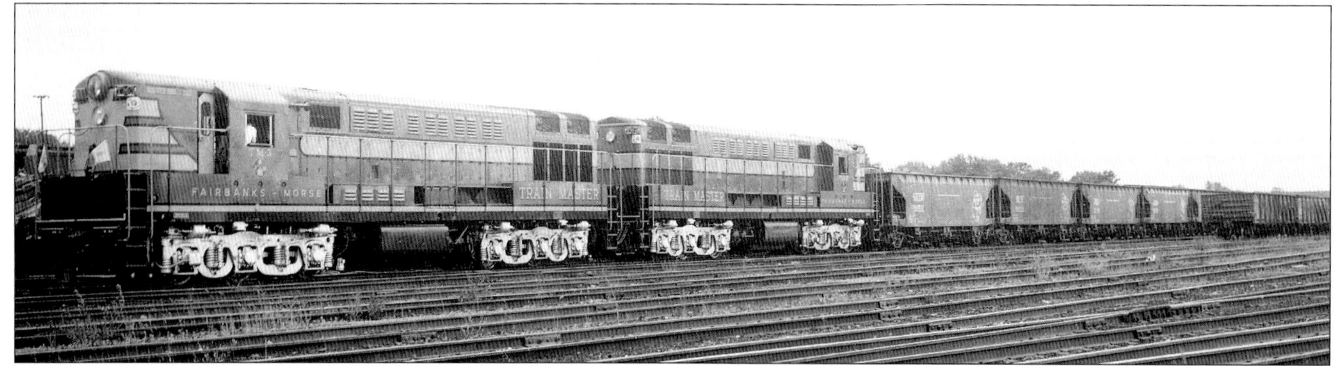

TM-3 and TM-4 prepare to depart Savannah, Ill., in May 1953 with an empty hopper train bound for the gravel pits at Beloit. Milwaukee Road gravel train service gave the Train Masters a thorough workout before heading out in demonstration and revenue runs.

lows [PRR employees] had told me that they did a good job, but they were not put in premier service."

Stacy also helped follow the TM-3 and TM-4 demonstrators on their trip through the West, and recalls being sent in 1953 to accompany them on the Southern Pacific Railroad. "The SP was a hot prospect," he said. "I rode out on passenger trains, caught up with them in Sparks, Nev. The Train Masters were working west on a freight; I got on at Sparks, went over the Donner Pass and down on the other side and wound up at San Jose (Calif.). The purpose was to test out in the San Jose-San Francisco suburban service, which later became the home of all the SP Train Masters.

"Everything went fairly normal, [except for] a problem with the governor of the engine. An incident happened when I wasn't on board. One of the Train Masters quit on one of the suburban trains but fortunately, it was between the two switches of a siding, so they just opened the switches and used the siding and the train didn't tie up the line. The Baldwin guys were getting unseated out there by the coming of the Train Master, 'cause they'd been selling a lot of six-motor equipment to the SP. SP was pretty partial to Baldwins, possibly due to the expertise of John F. Kirkland, who was a famous Baldwin salesman out there. Anyhow, Baldwin was ready to make big hay with that and then they realized that it was a Westinghouse failure in an auxiliary generator. Of course, Baldwin used Westinghouse, too, so they couldn't make much of it. I went out to the airport to meet the plane that was carrying the replacement [part] out from East Pittsburgh. So that was over in a few days and I came back on the *California Zephyr*."

Until many years later, Stacy was unaware that New York Central was a prospective customer that had even placed an order. When he heard about it, he compared notes with Robert Aldag, a retired FM sales engineering manager, who gave him the details. "The sales manager and salesman got involved in some kind of behind-the-door deal on the New York Central," Stacy said. "It was for eight Train Masters. [FM] started to build them for stock. All the locomotive builders did that at one time or another, they'd get a hot order they'd think was coming, they'd start building to keep the line going. That fell through and the SP [order] came along, so the units which were actually under construc-

tion became SP. [We] gave them some quick delivery. And, of course, they got the demonstrators, 3 and 4."

Stacy was along for the road testing of the first production Train Masters, those built for Delaware, Lackawanna & Western. To avoid snags in delivery, FM tested them in pairs, often with one painted and one unpainted unit. "That made the sequence come out," Stacy said. "One would be finished up without testing, go to the paint shop, and get painted, and then it'd be ready to go out, and the other one would be stationary-tested, factory-checked, and it would go out on the road with this one. Then this one would go to the paint shop to be painted. [It] fitted into the production schedule that way."

The first Train Masters equipped with General Electric electrical gear were those built for Virginian Railway, which came in 1954. The introduction of GE equipment simplified Stacy's job, which was that of being "sort of like a test foreman, so I really had to watch everything," he said. "The main purpose in a road test was to check transition control [of current to traction motors], like when you went from series control to parallel and back again, and the dynamic brake. You'd separate the units electrically and put one unit in dynamic braking, then you load the other one down like it was hauling a 10,000-ton train to make transition, and you gradually apply the brake. One routine was to put the dynamic brake on the rear unit, graduate it to let the head unit accelerate, go through transition, and as soon as he'd gone through transition, why, gradually apply the brake heavier and heavier and make it come back down again, and go through backward transition, as it's called. Then at the same time you could check the dynamic brake on the trailing

unit. The later [Train Masters] had automatic control, like 900 amps was the limit. Earlier units didn't have that (Westinghouse) but the GEs [did]. In other words, you'd get to 900 and you can't go any higher and burn something up. So you'd want to see if that was working properly.

"On the earlier units, you had to adjust just about everything when you went out on the test, but on the later units with the GE equipment, we got so that we could preset all the relays and resistors and stuff that governed that, and you could go out on the road and not have to touch a thing. They were very consistent. We weren't able to do that till we got the GE [equipment] – at least, we didn't find the combination that you could use with Westinghouse."

Stacy never made a trip to work with the Virginian Train Masters, but heard stories from his colleagues who did: "Of course, the N&W [in a 1959 merger] inherited all the Virginian Train Masters. But before that happened, they used to pull some interesting stunts. The Virginian would detour over the N&W, and they'd have a couple of Train Masters on a train, and those N&W guys would set the signals against them in a bad spot, you know, an interlocking or something like that, just to see if they could get the train started. And they did, every time, much to the consternation of the N&W, who didn't have a single diesel on the whole railroad at the time. They were proud of their big Mallets, as they should have been, because they were great engines."

Jersey Central also was an early buyer of the Train Master, Stacy noted. "We had delivered 1,600 horsepower [road-switcher units] before," he said,

Testing solo in Milwaukee Road freight service, TM-3 eventually was sold to Southern Pacific and spent most of its career hauling San Francisco-area commuters.

"so they [CNJ officials] were not new to FM, but the Train Masters of course were new to them – 2401-2407 were the numbers. They were a standard Train Master locomotive geared for 80-85 mpg and they were assigned to some of the clocker trains between Jersey City and Reading Terminal, the Reading clocker trains.

"The first Jersey Central locomotives were delivered around 1954, and then a second order came along. They were supposed to replace the famous Baldwin double-enders, which I had also helped build when I was at Baldwin. I always admired those engines – they were a good utilitarian [design], I mean, a double-end engine was what they needed up there. EMD E7s wouldn't have filled the bill operating-wise. So they were ready to get rid of them after a short life – I considered it a short life – and replace them with Train Masters.

"But they had one misgiving: They were unhappy with the riding quality of the Train Master truck. They ran a series of tests called the Jenkintown tests, which were done around Jenkintown (Pa., on the Reading). The tests convinced them that the truck was not satisfactory and [CNJ] came to us to see what could be done about it.

"It wasn't too hard to decide that the truck, which had a rigid bolster, could be built with a swing bolster, like most four-wheel trucks, EMD and Alco and a lot of the GSC-designed trucks. This truck was redesigned with a swing bolster, which went way up into the carbody, not too hard to do, so the next [CNJ Train Masters] were outfitted with the swing-bolster truck.

"The Jersey Central guys wanted to know if we could test that truck at high speed. Right away, we figured yes, because we had good relations with Milwaukee; all we had to do was line up a trip. We went from Beloit, with a light engine, to Sturtevant, then north [on the high-speed Chicago-Milwaukee *Hiawatha* route] to Milwaukee south city limits, to a town called Lake. That was the hottest piece of track on the system; it's still there as the Soo Line.

"And there was an 85 mph locomotive that we ran at 100 or so, just like the Milwaukee [passenger] trains did, and it ran just as smooth as could be. I don't know but what an original Train Master truck on a new locomotive could do just as well, we never tried one. But this one did just fine, and they were pretty well satisfied." Stacy rode the engine with George C. Wilms Jr., Jersey Central's assistant superintendent of motive power and rolling equipment; they made four round-trips over the 16-mile stretch between Sturtevant and Lake.

"In 1955, Fairbanks persuaded the Southern, which had bought six railcars in 1939 from St. Louis Car [with FM] opposed-piston engines in them. They had various kinds of experiences with it, but anyway, the famous Henry Taylor who had worked on the motorcars and was a mentor of Stanley Crane of the Southern who was later to become chairman of Conrail, was diesel superintendent at the time. And I got to work with him and his assistants. They had a very hot freight operation between Cincinnati and Chattanooga, and a lot of it was auto parts coming from Detroit. So they decided they'd cook up a combination locomotive that would have a Train Master in the middle and an EMD F7 on either end, and that gave the comfortable cab for the crew. The Train Master engineer's seat and cab layout weren't the comfort things that the F7s had. EMD engineer seats were like barber chairs; not so on the FM Train Master. So the Train Master just functioned as a B unit, so to speak.

"One time I boarded one of these outfits in Cincinnati, went right by the Cincinnati Union Terminal but of course didn't stop there, crossed the [Ohio] river, then ascended a rather steep grade, called Erlanger Hill, in Kentucky. And that was a tough hill to climb with the tonnages that they had. Well, we'd go up on dry rail and I could hardly believe what I saw. The train got slower and slower and slower and until it got down to 5 mph. That's much too slow for the amperage capacity of traction motors, even on a Train Master, but that's what we were doing and the ammeter was way over in the red zone. The TM, I suddenly realized, was doing all the work. The

FM engineers extensively tested the Virginian Train Masters, because they were the first TMs equipped with General Electric electrical gear instead of Westinghouse-built equipment. Virginian 807 (to be renumbered 50) and 52 sit in front of the Beloit depot in March 1954.

F7 in front was slipping, which is not all that surprising, but the F7 on the rear was also slipping, and that isn't supposed to happen. That EMD truck, the Blomberg truck, apparently it would cant. One pair of wheels would be almost off the rails, and that would lead to the slipping.

"We got over the grade, and away we went, and from then on I don't recall any grades that were as bad as Erlanger. But the tunnels were the meanest things – you couldn't ride the second unit without getting half asphyxiated. I learned to ride the first unit going through those tunnels. So that was one of the highlights of the Rat Hole Division. I don't know how much the Southern supervision rode that middle unit to see what was happening, but nothing was ever done about it, that I knew of, [because] the tonnages remained the same."

Stacy's particular job was to help measure the availability of air for engine intake in the CNO&TP's many tunnels. "The [FM] service people wanted somebody down there to look at the operation. We had a monometer, I know, and was supposed to read it as we went through the tunnel. We were trying to see whether there was any connection between the engine troubles that they had and the tunnel operation." A few years later, when Stacy returned, he found that many of the tunnels had been daylighted.

The combinations of EMD and FM locomotives on the Southern required a solution to the problem of incompatible dynamic-brake systems. Stacy relates: "We invented at FM – we were the first to have it – what was called universal dynamic brake control. EMD controlled their dynamic brake with a field loop, it went down through all the units and then back again, a loop circuit. General Electric did it with a potential circuit, varying the potential, and they were not compatible directly. So when we got to a rig like this, we fixed it so you could use either one. Our Train Master unit responded to a field loop signal, but when it was running by itself or another GE, we used the potential circuit. They were actually wired in parallel off the same controllers so that you could run it like that. We got so that we called it universal dynamic brake control."

Another service call Stacy made was a trip to Canadian Pacific's Nelson, British Columbia, shops. "That's the Kettle Valley Railroad – they had some

The standard locomotive lashup on Southern Railway's Cincinnati-to-Chattanooga main line was an F7-TM-F7 lashup, as seen at Danville, Ky., in the summer of 1955.

terrible grades up there. That is the curviest railroad there was in North America. They were running C-Lines on the passenger trains. Train Masters were on the freights. They said that they were burning up traction motors. They checked the air supply to the traction motors and found it was abnormally low. Well, the first thing you wonder is, how did they check it? So we made a check at Beloit in the prescribed manner, got a [monometer] pressure [reading] of 5-1/2 inches of water. Out in Nelson, they said they were only getting something like 3.6, which is not enough air flow.

You have a monometer, you hook it up to the air stream and it exerts pressure which translates into the differential between one tube and another and that's the way you measure air supply. So my job was to go out to Nelson - they had a [monometer test] rig out there – and to make sure it was exactly the same as the one in Beloit. Again, 3.6. Couldn't account for it. I came back with no answer 'cause I couldn't figure it out.

"I don't know how long it took, but somebody finally found out that Model 752 traction motors in the Canadian locomotives were manufactured at Peterborough, Ontario, by GE, and they did not have the same internal dimensions that the American Erie-built traction motors did. The result is that you could put the same blower [on both types] and on one, you'd get low air pressure and on the other you wouldn't. And that was the answer. It was within the traction motor itself – it was a resistance inside the motor. I didn't know things like that were allowed to happen. So how did they fix it? I don't know to this day what they did, whether they changed the motors. You'd think it would've affected them on other locomotives. Alcos, for instance, had the same traction motors."

In 1956, Stacy said, came the beginning of FM's departure from the locomotive business (FM also made pumps, scales, and diesel engines for non-railroad use), but it had nothing to do with technology, product quality, or service: "There was something akin to what you see a lot of today, and that was the corporate raid. The Morse brothers had split up, which was the root cause of this thing, and some guy by the name of Silberstein came along and got one branch of the Morses on his side and started to raid the company – buy stock, tried to get control. It was Bob Morse and Charlie Morse, Chuck Morse as we knew him, [he] was on the wrong side, I remember that. [He] threw in with Silberstein and that sort of overbalanced Bob Morse, I guess, he couldn't hold onto it. Tried, with his allies, but the other side wound up winning."

The battle for control of the company turned into a proxy fight, which scared off at least one big order and permanently put FM out of the locomotive-building business.

"The Illinois Central was the big one," said Stacy. "That was [a potential order for] 75 locomotives. We went down there and demonstrated, I suppose with the 1 and 2. They were impressed, they were really impressed, and they were all set. And then the proxy fight hit. And of course that killed all of that, that was the end of the business.

"It affected everything, it wrecked morale. Everybody knew what was going on. [There was] no money available for lots of things – expansion, even maintenance money. Every cent they could get their hands on was taken to buy stock and hold onto it to keep the opposition. The stock went from about 25 to 65 overnight. There was one of the guys in the service department, Wayne Sidmyer, and he had a bunch of that stock, so did some others. [Management] used to meet him at the gate every morning and see if he held onto the stock, because the opposition was buying all the stock it could get hold of. And of course anybody that had a bunch of stock that went up 40 points could make a big profit if he sold. Some of them did, I guess, there were a lot of them in that category. If you sold it you were as good as fired.

"Top management changed as, I guess, they became new owners, took over, [and] then the company became part of Colt Industries. I figured that was the end, the end of the locomotive business, and it was. It tapered off till there finally was nothing but parts business, which was good, but not enough to support everything like the live locomotives used to.

"About then I started looking for another job. I wanted to stay in the railroad end if I could, and I thought I had a sure chance with the Budd Co. in Philadelphia. It didn't work out – they were in bad times, too. By '57, I think the ($25 million, 270-car) Philadelphia Frankford El car job was the only thing that saved them from extinction at the time. They did well with that, but they didn't have a whole lot else. I wasn't going to jump off unless I had a good place to land. I didn't know whether I could hang on without them ditching me."

But while the company retrenched, it continued to support Train Master customers. In 1959, Stacy made a service call on the Wabash, which turned out to be both memorable and unmemorable. "A fellow by the name of McGregor who had been an Alco man [was] in a supervisory diesel position on the Wabash, and I worked with him briefly. I don't remember what we went over, but I didn't ride any of the units over the road. The only thing I did, I had to go out in the middle of nowhere in Missouri at some crossing and pick him up off the locomotive, and drove to Decatur (Ill.) in the worst ice storm, which lasted for three days. I had to drive back to Beloit in that. Awful."

Stacy ended up taking a job as a transit engineer for the City of Philadelphia in November 1964, and didn't miss a day of work between the two employers. "My last day at Beloit was on a Friday and I worked some overtime there getting a switch or wiring job done. I rode my bike home around 6:30 that night, and then Saturday caught the train in Janesville and headed for Philadelphia. I stayed overnight with my friend Ken Darling in Wayne, Pa., then took the train to work (Monday morning)." In that job, he worked with equipment, scheduling, and planning (he did some early work on the Center City Commuter Tunnel, which was not built until the 1980s) and dealt daily with the Pennsylvania (later Penn Central and Conrail) and Reading railroads and with the Southeastern Pennsylvania Transportation Authority.

Even then, he ran into an old friend in the form of a Jersey Central Train Master that was still assigned to a Reading-CNJ Philadelphia-Jersey City passenger run. "It came out of Jersey City in the morning and came into Philly about 9 o'clock. It came in as a commuter train on the Philadelphia end, and it left here at 4:52 p.m. and went back to Jersey City. So I got to see that for a while and then it turned out that I was sent up to check the ridership of that [train] north

Bob Stacy

of West Trenton, and it was good. I got to know the conductor and he said, 'Well that looks good, but you know they're all traveling on New Jersey state passes,' of which there were a prolific number, [state employees] going to work in Trenton, I guess.

"So the train [revenue] wasn't doing too well north of West Trenton and it eventually came off and was replaced by an MU on this end. So that was the end of the Train Master. But I used to take a nostalgic look at that Train Master. This was 1964, now, in Reading Terminal. It used to sit there with the engine shut down, which is something you could do. I suppose they left it running on the coldest days. Every time I saw it, it was shut down, up against the bumping post."

Of the Train Master fleet as a whole, Stacy had no strong recollections of his last official contacts with it as an FM employee, but offered this plain observation: "Like old soldiers, they seemed to fade away." – *Dan Cupper*

Just before departing on their first demonstration tour, TM-2 and TM-1 sit on the Fairbanks-Morse plant lead at Beloit. Although the Train Master was successful, the builder's minority status when compared to the number of locomotives sold by the other two major locomotive builders put it at a disadvantage that it never overcame.

FM/David R. Sweetland collection

A longtime Fairbanks-Morse customer, Pennsylvania Railroad sampled the Train Master model with an order for nine units, including 8706, photographed at Beloit in September 1956.

Pennsylvania

Number of Units: 9

Road Numbers: 8699-8707

Year Built: 1956

When the Pennsylvania Railroad placed an order for nine Train Masters in 1956, the event turned out to be notable in two ways – it was Fairbanks-Morse's 200th locomotive order and it would be PRR's last order for FM locomotives.

By the time the Train Master was introduced, the tide of dieselization had temporarily subsided on the Pennsy and the carrier went about fine-tuning specific locomotive assignments. Although the Train Master had been introduced in 1953, it wasn't until 1955 that PRR ordered 2,400-horsepower diesels, Alco DL-600A/RSD-7 models. The units were purchased to work in helper service between Altoona and Gallitzin, Pa., assisting passenger and mail trains on the 11-mile grade. They were meant to replace F3 A-B-A sets in helper service, releasing the cab units to road freight service where their 4,500

horsepower could be put to better use. The Alco units proved to be less than satisfactory and probably tainted PRR's operating and mechanical departments on the merits of a single-engine high-horsepower locomotive, because the DL600B/RSD-15 models that arrived the following year were purchased specifically for hump and transfer work.

But the 2,400-horsepower unit did find a place on the system, because in addition to the RSD-15s, PRR ordered nine Train Masters in 1956. By that time, Pennsy had assembled a sizable Fairbanks-Morse roster – 48 Erie-builts, 24 C-Liners, 55 H10-44s, 16 H12-44s, 38 H20-44s, and 10 H16-44s – the railroad was by far FM's best customer.

left, *Equipped with trainphone equipment for road service, Pennsy's Train Masters carried the distinctive rooftop antenna, as seen in this rear view of 8706. Note the fuel filler spout in the end of the short hood.*

FM/Robert Watson collection

Don Pope

Prior to renumbering, 8703 awaits its next assignment at Columbus, Ohio, in the mid-1960s. Pennsy's Train Masters were equipped with General Electric rotating equipment – six Model 752-E1 traction motors and a Model GT-567 main generator.

Delivered in August and September 1956, they came with Phase II carbodies and Pennsy's distinctive trainphone antenna. Assigned PRR classification FS-24m (Fairbanks-Morse, Switching, 2,400 horsepower, and multiple-unit equipped), the units weighed 376,000 pounds, allowing a starting tractive effort of 94,000 pounds at 25 percent adhesion, approximately that of one of PRR's I1s-class 2-10-0 steam engines. PRR opted for three fuel tanks with a combined capacity of 4,200 gallons. Equipped with a 74:18 gear ratio, the units had a top speed of 66 mph. With all of the PRR options, each Train Master cost $254,602.

The units, numbered 8699-8707, were assigned to the same duties as their Alco counterparts. Four units, 8699-8702, went to the Buckeye Region where they were assigned to Columbus, Ohio, for yard and heavy transfer service, while 8703-8707 were assigned to the Pittsburgh Region where two were assigned to East Altoona, Pa., and three to the 28th Street enginehouse in Pittsburgh. Four of these, 8703-8706, were operated in pairs as West Slope pushers out of Conemaugh (Johnstown), Pa., while 8707 worked the Brilliant Branch in helper service.

During the 1950s and early 1960s, the Train Masters remained in their original assigned roles, with pairs of TMs replacing A-B-A sets of F3s in helper service. Crews were impressed with their pulling power at low speeds but commented on their sand consumption. By 1960, the four TMs assigned to helper service were reassigned to Conemaugh, while the 8707 remained at 28th Street.

By mid-1962, all nine H24-66s congregated at Columbus, Ohio. With the upcoming merger with New York Central, the PRR Train Masters were renumbered to 6700-6708 in 1966. By 1969, most of the TMs were stored out-of-service at East Altoona, Pa., waiting for their 15-year equipment trusts to expire. The exception was 6700, which had been assigned to Chicago's 59th Street Yard hump in 1967. In 1970, the unit was renumbered to 6799 to clear a number block for General Electric U23Cs being delivered. Retired in May 1970, the unit went to the storage line at East Altoona still carrying PRR markings. ★

Richard O. Adams

Wearing its pre-merger road number, an oil-covered 6701 was working at Columbus, Ohio, on September 23, 1967. The four H24-66s assigned to Columbus spent their entire careers working in the Buckeye state before being stored at Altoona waiting for their equipment trusts to expire.

The Garys

above, *The large radiator cooling fans common to the Train Master are prominent in this view of 6706 at Columbus, Ohio, on June 13, 1967. The radiators and cooling system held 250 gallons of water.*

right, *This down-on view of 6706 at 20th Street Yard in Columbus, Ohio, on November 26, 1966, shows the recess designed into the short hood to accommodate the cab door; the rain gutter on the cab roof; the large dynamic brake cooling fan; and the exhaust stack spark arrestors.*

below, *Wearing Pennsy's simplified lettering scheme – six keystones – 6708 was equipped with a canvas cab awning. Columbus, Ohio, September 23, 1967.*

J. David Ingles/Louis A. Marre collection

Richard O. Adams

George Melvin

above, **With its numberboards removed and a "0" painted before the road number, 6703 awaits the scrapper's torch at East Altoona, Pa., on March 5, 1970. The addition of a zero to the road number of a retired or soon-to-be retired locomotive was a PRR tradition dating from steam days.**

right, **Although the "COL" stencils (for Columbus, Ohio) appear to be fresh, 6701 was stored out-of-service at East Altoona, Pa., in this September 18, 1969, photo. Even though a walkway was provided at the ends of the PRR Train Masters, a solid handrail was applied – note the walkway chain.**

C. N. Shankweiler/Kenneth M. Ardinger collection

Jim Claflin/Kenneth M. Ardinger collection

With only a few months of service remaining, former PRR 8700 now wears its third road number, after first being renumbered 6700 for the Penn Central merger. Note the mechanical speedometer cable attached to the center axle on the rear truck. Chicago, March 20, 1970.

Leading a Pottsville-bound passenger train at Linfield, Pa., on August 18, 1956, is steam-generator-equipped H24-66 863, one of six dual-service Train Masters acquired by Reading in 1955 to complete dieselization of the anthracite carrier.

Martin S. Zak

Reading

Number of Units: 17

Road Numbers: 800-808, 860-867

Years Built: 1953, 1955, 1956

Like other eastern carriers, the Reading Co. saw the Train Master as a way to finish its transition from steam to diesel power with the purchase of a minimum number of units. Although it was initially interested in Fairbanks-Morse's 1,600-horsepower H16-44 road-switcher model, it was the TM's 2,400 horsepower that impressed Reading's management – two Train Masters were equivalent to three 1,600- or 1,750-horsepower models such as those offered by the major builders. The railroad's interest was based solely on the Train Master's advertised specifications, since Reading issued a purchase order for four H24-66s in April 1953 – two months prior to the release of TM-1 and TM-2 for demonstration tours.

Reading finally began testing the two demonstrators on July 1, 1953. Delivered to Reading at Port Reading, N.J., by the Jersey Central after the

One of two H24-66s purchased for hump service at Reading's largest classification yard, Rutherford, 800 carries an "RU" stencil on its pilot denoting the identification of its maintenance base. Rutherford, Pa., March 25, 1956.

John D. Hahn, Jr.

As delivered, the long hood on all of Reading's Train Masters was designated as the front. Unit 804 was part of a five-unit order for freight-equipped H24-66s – dynamic brakes and hump controls were specified for these units. Philadelphia, March 17, 1957.

conclusion of the Railway Supply Manufacturers' Association show in Atlantic City, N.J., the units operated in a variety of assignments across the system for seven days. At the conclusion of the tests, Reading officials felt that the Train Master was equivalent to the company's massive 219-ton 2-10-2 K-class steam locomotive. On the carrier's steepest grades, a pair of Train Masters was an even match for Reading's three-unit sets of Alco FA-1/FB-1s or EMD FTs/F3s.

While Reading's first TM order was for two units for freight service and two for passenger duties, the carrier immediately placed a second order for five

more freight units. The units were numbered accordingly – 800-806 were equipped with dynamic brakes and hump controls for freight work, while 860 and 861 carried a steam generator and dual controls, but lacked dynamic brakes for passenger assignments. All of the units, regardless of assignment, were given 68:15 gearing. The freight units were equipped with remote boiler controls and steam train lines so they could be mated to the 860-series units in passenger service. The 800-series units were assigned Reading class RS-4 (Road Switcher type, fourth type purchased) and cost $237,058 each, while

the 860-series units carried RS-4b class (the b standing for steam-boiler-equipped) and cost $241,118 each. On both freight and passenger units, the long hood was designated as front. They came painted in Reading's standard dark olive green paint scheme with dulux gold or dark yellow lettering. The trucks and fuel tank were painted black.

The Train Master was designed with three tanks – one for fuel (1,800 gallons) and two that could be used for either fuel or water for steam boiler supply (1,000 and 1,400 gallons capacity, respectively). Reading opted to equip its freight units with all three tanks for a total fuel capac-

Although both H24-66s are passenger-equipped, 861, built in 1953, features a Phase Ib carbody without dynamic brakes (no open grid just ahead of the cab), while 863, built in 1955, displays a Phase II design with dynamic brakes. When delivered, the 860-series TMs were assigned to Reading's Green Street (Philadelphia) enginehouse for maintenance. Saucon Creek, Pa., December 26, 1954.

Still carrying its St. Clair, Pa., assignment (note the "St.C" stencil on the pilot), 802 was working in road freight service when it was photographed at the Rutherford, Pa., engine servicing facility on April 3, 1961. The Reading's Train Masters carried two Wabco H-6 model horns, one recessed into the short hood, and the other mounted on the engineer's side of the long hood. The nose-mounted horn was unique to Reading's Train Masters.

John D. Hahn, Jr.

This view of 802 at Philadelphia in February 1964 shows the rain-gutter strip applied to the cab roof on all of Reading's locomotives.

R. S. Short/The Garys collection

Reading was one of two railroads to order its Train Masters with 4,200-gallon fuel capacity. This required three separate tanks, but all were filled through one spout located on the front of the short hood. Rutherford, Pa., April 23, 1962.

John D. Hahn, Jr.

The Train Masters weren't the first Fairbanks-Morse-powered locomotives on Reading's roster. In 1939, the carrier took delivery of a 600-horsepower center-cab locomotive built by St. Louis Car Co. and powered by a pair of eight-cylinder 300-horsepower in-line diesel engines built by FM. The trouble-prone unit was traded by Reading on its first TM order. Freshly washed 805 gleams in the fall light at Philadelphia on October 25, 1964.

Martin S. Zak

Recently reassigned to Rutherford Yard, 860 would soon lose its steam generator and RS-4b classification. Rutherford, Pa., March 27, 1964.

John D. Hahn, Jr.

Demonstrating its dual-service capability, 863 prepares to depart Harrisburg, Pa., on August 6, 1962, with eastbound Train 194. The following year would see the end of Reading passenger-train service in Harrisburg.

Dave Engman/The Garys collection

ity of 4,200 gallons, while the passenger units split the use of the three tanks – one 1,800-gallon tank (mounted below the frame between the trucks) was for fuel and two tanks (mounted inside the carbody) were allocated for 2,400 gallons of water. The freight units weighed 385,900 pounds and the passenger units weighed 386,700 pounds.

The first two Train Masters, 800 and 801, arrived at Rutherford, Pa. (Reading's Harrisburg-area yard), on October 4, 1953, while 860 and 861 arrived on October 16. The five remaining freight units arrived on October 28, 1953. The nine Train Masters were joined by 42 other diesels (Alco RS-3s, EMD GP7s, and Baldwin AS16s) in 1953. The impact of these diesel acquisitions on Reading's steam fleet was clear, and stark statistics told the story. At the start of 1953, Reading operated 135 steam locomotives – by the end of the year, that number had dropped to 73 locomotives.

The diesel locomotive fleet remained constant until 1955, when six dual-service Train Masters were delivered in November and December. Numbered 862-867, they were equipped with dual controls (862-865 only), steam generators, dynamic brakes, and hump control, and the long hood was designated as front. They were the last TMs built with Westinghouse electrical equipment. Like 860 and 861, this group was assigned Reading class RS-4b. Units 862-865 weighed 388,400 pounds, while 866 and 867 weighed 386,400 pounds. The delivery of these units, for all practical purposes, completed dieselization of the Reading lines, although 40 of the carrier's

Leading a Rutherford, Pa.-to-Hagerstown, Md., mixed freight at Carlisle Junction, Pa., on July 25, 1963, are 862 and EMD GP7 631. The mixture of hopper cars with general merchandise cars was typical Reading practice.

The ground shakes as a pair of Train Masters, with a Pennsylvania Railroad L1s-class 2-8-2 helper, leads an eastbound coal train through Williamsport, Md., on June 17, 1956. The Baltimore & Ohio-originated train was interchanged at Cumbo Yard in Martinsburg, W. Va., and operated via PRR trackage rights to Shippensburg, Pa., where it continued on Reading trackage to Rutherford and points east.

most modern steam engines remained on the roster to handle traffic surges.

The final first-generation road diesel purchase for Reading was a pair of Train Masters delivered in December 1956. Unlike all the previous Reading TM orders, which came with Westinghouse electric equipment, this pair of units was built with General Electric rotating gear. Long-haul passenger service was being discontinued, so there was no need to equip these units with steam generators. They were assigned road numbers 807 and 808 and were built with hump controls and steam train lines. Like 800-806, they carried 4,200 gallons of fuel, but with the GE equipment, they rode on 40-inch

Late in December 1956, Reading acquired its last two Train Masters, 807 and 808. These featured Phase II carbodies and were the only Reading units to be delivered with General Electric electrical equipment. Unit 807 was photographed at Philadelphia in March 1963, looking clean with its fresh coat of paint.

wheelsets with 74:18 gearing (all of Reading's Westinghouse-equipped TMs rode on 42-inch wheelsets with 68:15 gearing). They weighed 388,400 pounds.

Initially, two of the freight-equipped TMs, 800 and 801, were assigned to hump service at Rutherford, Pa., and five units, 802-806, were assigned to anthracite mine service and based at St. Clair, Pa. During the week, one unit was based at St. Clair working the scale track, and one unit was based at Tamaqua, Pa., with the remaining units working mine turns. On weekends, the Train Masters made freight runs to Reading, returning to St. Clair in time for Monday morning assignments.

The passenger-equipped H24-66s replaced pairs of RS-3s and AS16s on Bethlehem Branch passenger trains and on Philadelphia-Shamokin, Pa., runs (Trains *6/97*, the *King Coal*). As hard-coal production dwindled in the late 1950s, many of the Train Masters were reassigned to road freight service between Allentown, Pa., and Rutherford. Equipped with standard MU connections and field loop dynamic braking to multiple with EMD power, they were typically mated to a 1,500-horsepower EMD cab unit or road switcher, or operated in 4,800-horsepower pairs.

Shortly after delivery, the carrier experienced vibration problems with the tri-mount truck at high speed. Reading, in conjunction with Jersey Central,

H24-66 808 hustles a mixed freight through Bridgeport, Pa., on September 2, 1964. These Phase II units featured the larger rubber-gasketed numberboards.

With the delivery of second-generation models, the Train Masters were gradually demoted to yard service. But when it was caught short of power, Reading pressed the aging workhorses into road service, frequently mated to EMD GP30s and GP35s, and, as in this view at Rutherford, Pa., on November 26, 1965, Alco C-424s.

which was also experiencing the same problems with its Train Masters, conducted a series of high-speed tests called the Jenkintown tests (named for the line on which the tests were conducted). As a result, FM and General Steel Castings Corp. designed a swing-bolster version of the tri-mount truck, which was retrofitted to Reading's 1953-built H24-66s and came as standard on the carrier's 1955 and 1956 TM orders.

Around 1958, Reading conducted a series of tests resulting from crew complaints about engine exhaust fumes entering the TM cabs. As a result, the short hood was redesignated as front on the H24-66s, beginning in late 1958.

With the delivery of second-generation diesels in the early 1960s, the Train Masters continued to work in road service, but now were mated to one of the carrier's new EMD GP30/GP35, or Alco C-424, models. Two units, 801 and 867, were repainted into the carrier's green and yellow scheme, which had been introduced with the delivery of its GP30s in 1962. Unit 867 emerged in late 1965, and in January 1966, 801 was released wearing this colorful attire.

With the delivery of new Budd RDC cars in 1962 (additional secondhand cars were purchased in 1965-66), Reading's 900-series FP7s were assigned to cover the few remaining locomotive-hauled passenger trains. As a result, steam generators were removed from the 860-series Train Masters in 1964-65. Correspondingly, Reading reclassified the units from RS-4b to RS-4.

Only two TMs were repainted into Reading's "modern" image before the company decided that the units had reached the end of their economic usefulness and no more major capital expenditures would be made on the RS-4s. Unit 801 is at Bethlehem, Pa., on April 17, 1966.

In late 1967, nine of the remaining active H24-66s were renumbered, including former 808, which was renumbered 203 on October 17, 1967. A heavy coat of dirt accumulated on the Train Masters during their final years of service – attributable to both their assignment, yard and hump service, and to their lowly status on the maintenance chart.

above, *Units 265 and 261 await their turn at Rutherford's engine servicing facilities on November 25, 1967. Note the radio antenna on the short hood of former 867.* below, *By the late 1960s, Reading assigned its remaining TMs to Rutherford, where they were readily accessible to maintenance personnel. Reading 201 is at Rutherford, Pa., on November 24, 1967.*

In 1967, the fleet had been reduced to 15 units – 804 was burned at Gettysburg, Pa., in 1966 and stored at Reading, while 807 was retired in mid-1967 and sold for scrap – but many of these units were in storage. To avoid number conflicts with Reading's MU commuter cars, nine of the remaining active units were renumbered into the 200 series, with renumbering taking place between August and October 1967. With new high-horsepower second-generation units moving Reading's road trains, the remaining Train Masters were assigned to hump and yard service at Rutherford.

The renumbered TMs carried their new numbers for only three years – in April 1970, 203 and 260 were retired, while 261-265, 861, and 866 left the roster in June. ★

Both photos this page, Robert M. Stacy

Awaiting its next run, an F7/H24-66/F7 lashup poses in front of the Danville, Ky., station during the summer of 1955. Built by EMD in June 1951, F7A 4261 was retrofitted with dynamic brakes in July 1955 for operation on CNO&TP with similarly equipped Train Masters.

Southern

Number of Units: 5

Road Numbers: 6300-6304

Year Built: 1955

A carrier that was always on the cutting edge of diesel technology, Southern Railway proved the fact again by purchasing the Train Master. But the southeastern carrier's dieselization roots were firmly planted in La Grange – its first road freight diesels were EMC/EMD FTs, while the carrier's first passenger diesels were slant-nosed E units. During the war, Southern acquired a fleet of 68 FTs to move mainline freights, while a sizable number of Alco and EMD end-cab switchers handled yard duties. At the end of World War II, the carrier began to replace steam locomotives with a vengeance – between 1946 and 1952, 254 EMD F2, F3, F7, and FP7 models arrived. Southern also recognized the versatility of the road switcher carbody and placed 32 Alco RS-2s placed into service in 1948-49, followed by 98 GP7s in 1950.

But in the early days of dieselization, Southern did look at and purchase products from the other builders, though in much smaller quantities. The lone Baldwin VO660 and VO1000 models purchased in 1941 were joined by five DS4-4-1000s in 1948. The first Fairbanks-Morse locomotives, five H15-44s, were delivered in 1949. These were followed by 10 H16-44s in 1951 and four H12-44s in 1953. But on a roster that numbered 584 diesels (on December 31, 1953), 19 FM products certainly constituted a minority. With dieselization completed in June of that year, it was obvious that the Train Master, no matter how successful, would have difficulty finding a home on the Southern roster – but the

This closeup view of 6302 and F7A 4261 shows the TM's low end platforms and modified radiator cooling fan openings, which resembles EMD's Tunnel Motor design introduced nearly 20 years later in 1972.

By the 1960s, the Train Masters were still used as booster units, but now mated to SD24s. Here SD24s 6325 and 2510 flank H24-66 6301 at Louisville, Ky., on July 5, 1964.

bottom, *Modified with stack extensions, roof-mounted ice breakers (to protect the horns), and walkway lights, 6301 is at Chattanooga, Tenn., on February 27, 1965.*

following year brought a pleasant surprise to the FM sales department.

In an effort to move freight more efficiently on Southern's most difficult operating division, the 338-mile Cincinnati, New Orleans & Texas Pacific Railway between Cincinnati and Chattanooga, Tenn., the carrier took a close look at the Train Master, and in 1954, placed an order for five of the 2,400-horsepower beasts.

Controlling interest in CNO&TP was acquired in 1895 by Southern, which managed the line through three operating districts. Stiff grades and numerous tunnels on the Second District earned it the nickname Rat Hole Division. Its rolling profile and contributions to the

H24-66 6300 and H16-44 2147 were nearing the end of their Southern careers in this February 27, 1965, photograph at Chattanooga, Tenn. Later that year, all five H24-66s and nearly all of the remaining H12-44s, H15-44s, and H16-44s were traded to EMD for a 100-unit order of SD35s. Although the initial Train Master order spawned no additional orders, it did prove that the C-C truck locomotive was well-suited for rugged CNO&TP lines.

bottom line – more than 10 percent of the yearly gross operating revenues – demanded the highest horsepower locomotives on Southern's roster. EMC's FT demonstrators tested on the Rat Hole, and later became permanent fixtures on the line. By the early 1950s, the FTs were superseded by 1,500-horsepower F3s, and later, 1,500-horsepower F7s. Typically operated in four- and five-unit sets, the cab units ruled the CNO&TP until 1960, although help arrived in 1955. Southern, always looking for ways to improve speed over the Rathole, took a close look at the Train Master's specifications and liked what it saw.

Delivered in May and June 1955, the units were built with Phase II carbodies, and to better mate with the cab units, were built with low end platforms. Bause of the grades on the Rat Hole, the units were equipped with dynamic brakes. Weighing 362,900 pounds fully provisioned, they were the lightest Train Masters built, but were the largest diesels on the Southern until EMD SD24s arrived in 1959. The units were painted in the carrier's green scheme with CNO&TP sublettering denoting ownership. A five-chime horn was mounted on the cab roof, and the bell was mounted on the front of the unit – the end of the short hood.

On delivery, each TM was mated with a pair of F7 cab units, with the H24-66 serving as a booster unit. Al-

most immediately, Southern began to experience overheating problems with the Train Masters due to the number of tunnels on the Rat Hole. As built, CNO&TP featured 27 tunnels, although by 1950, the number had been reduced to 13. To improve engine cooling, Southern added radiator fan shrouds to the TMs in the fall of 1955. By the spring of 1956, louvers similar in appearance to those used on F units were cut into the H24-66s' engine compartment doors for additional cooling. This must have solved the problem, as the Train Masters and their F-unit mates continued to run between Cincinnati and Chattanooga for the next four years.

The first major assignment change for the TMs came in 1959, when delivery

of 48 2,400-horsepower EMD SD24s began. With their extra horsepower and improved dynamic braking, three-unit sets of SD24s began to rule the main line, with the Train Masters continuing to serve as booster units, but now mated to a pair of SD24s. The SD24s also introduced a new paint scheme to Southern, with somber black replacing bright green as the predominant carbody color. Eventually, all five H24-66s were repainted into this scheme.

By the mid-1960s, their minority status on the roster began to work against them and they were frequently seen in the Chattanooga shops awaiting repairs. Retired in 1965, the TMs were traded to EMD on June 23, 1965, on an order for SD35 road freight units. ★

Southern 6303 is at Chattanooga, Tenn., on February 27, 1965.

Among the most photographed diesels on Southern Pacific, the Train Masters were a fixture on San Francisco-area "commute" trains (an SP term) for nearly two decades. In this view at Bayshore, Calif., in 1958, former TM-3 roars south with Train 124.

Jim Shaw

Southern Pacific

Number of Units: 16

Road Numbers: 4800-4815

Years Built: 1953, 1954

Southern Pacific first became an FM customer in 1952, ordering six H12-44 end-cab switchers. With a fleet of EMD, Alco, and Baldwin switchers already in service, it seemed out of character for SP to purchase opposed-piston-powered locomotives at this late stage of its dieselization. One reason could have been FM's ability to meet SP's delivery schedule. EMD was experiencing a backlog of orders in the wake of a national rush to dieselize, and

SP, like other railroads trying to eliminate steam power as quickly as possible, turned to Baldwin, Alco, and FM, all of which promised faster delivery.

This initial order was followed by a 10-unit order for H12-44s the next year, as Fairbanks-Morse began to established a foothold with a major western carrier, something it had failed to do with its earlier models. Although both Santa Fe and Union Pacific sampled FM's 2,000-horsepower Erie-built

model, FM failed to land any major orders, as both carriers experienced less-than-satisfactory performance. FM looked to Southern Pacific for the opportunity to capture a portion of the burgeoning road-switcher market, while SP was seeking a high-horsepower dual-service locomotive to replace Mt-class Mountain and GS-class Northern steam engines in the San Francisco commuter pool, as well as to move symbol freights at track speed with fewer locomotives.

FM

This builder's portrait of 4812 shows the typical Phase Ib carbody that was common to SP 4802-4815. SP opted for a five-chime horn (in place of the standard pair of single-chime horns) mounted on the short hood, which was designated as front on all SP Train Masters.

On October 12, 1959, SP 4810 prepares to back off the Mission Bay (Calif.) passenger roundhouse turntable at San Francisco. Both ends of SP's TMs were equipped with train number indicators and safety light for bidirectional operation. Note how the silver-painted areas were prone to fading.

Southern Pacific's first experience with the Train Master occurred in the fall of 1953 when Santa Fe delivered TM-3 and TM-4 to SP at Los Angeles. After a series of freight and passenger runs in the Los Angeles area, the units were shipped to San Francisco, where they operated both singly and paired in San Jose commuter service during October. In November, the units headed back to southern California and entered freight service between Los Angeles and El Paso, Texas. The tests were a success; SP purchased the pair in November 1953 and also placed an order for 14 more units.

SP's El Paso shop repainted TM-3 and TM-4 into the road's "black widow" painting arrangement in early December 1953 and numbered them 4800 and 4801. These were the first road switchers painted in the black widow scheme, but instead of painting the trucks black to match SP practice, the shop left them in their original silver-painted color. While TM-3 was originally equipped with a steam generator, TM-4 was not. Shortly after its purchase, FM supplied the equipment to retrofit TM-4 with a steam generator to match TM-3 and the 14 units on order.

Strangely, almost immediately after SP placed its order, delivery began. In December, 4802-4807 were delivered to El Paso, followed by 4808 and 4809 in January 1954, and 4810 and 4811 in February. The balance of the

order arrived in March. The quick delivery schedule stemmed from the fact that FM had eight TMs ordered by New York Central, but later canceled, in various stages of construction on the Beloit factory floor. NYC 4604-4607 became SP 4802-4805 and NYC 4600-4603 emerged as SP 4810-4813.

Subtle exterior differences existed between the various groups of SP Train Masters, owing to the "NYC heritage" of several units. Units 4802-4809 were delivered without SP-style train-number indicators, while 4810-4815 were equipped with them. The former demonstrators were assigned SP class DF-500

Train Master 4812 and SD9 5365 await assignments at Mission Bay enginehouse on November 12, 1959. Both units were assigned to the commuter pool – note the "SF/COM" lettering between the SP class notation, DF-501, and weight, 379 (for 379,000 pounds).

Former TM-4 rolls a commuter train through Brisbane, Calif., in 1958. Built without a steam generator for its demonstration tour, TM-4 was rebuilt by SP with a generator, using parts supplied by FM. Included were new ventilation panels for the short hood.

Jim Shaw

The victor and the vanquished – a string of retired steam engines serves as a backdrop for H24-66 4802 and Train 114, a four -car commuter train consisting of a baggage car, an RPO, and a pair of gallery cars. Bayshore, Calif., 1958.

Jim Shaw

The Garys

Laying over at San Jose, Calif., in the late 1950s, 4802 displays the train indicator board style common to 4802-4809. Note the small boxes mounted below the indicators for the removable train numbers, and the addition of a small plow-type pilot.

and weighed 382,100 pounds, while 4803-4815 carried SP designation DF-501 and weighed 379,320 pounds. All SP TMs carried 2,400 gallons of water, while all but 4806-4809 carried 1,800 gallons of fuel – 4806-4809 carried 1,950-gallon fuel tanks. All the units were built with Westinghouse electrical equipment, but SP extensively modified the TM's electrical control equipment, yet retained the Westinghouse main generator and traction motors. Although they were delivered with dynamic brake equipment, this system was deactivated in the early 1960s.

Like the FM switchers that preceded them, the Train Masters were assigned to the Tucson and Rio Grande Divisions, with El Paso designated as their maintenance base. They were assigned to both freight and passenger service, but this assignment didn't last long as the harsh climate of West Texas and New Mexico, with its sandstorms, played havoc with the Train Masters' air filter system. This, along with the fact that incompatibilities existed between the FMs and SP's sizable fleet of EMD locomotives, caused SP to look for another assignment for the 16 H24-66s. In August 1954, the two TMs were temporarily reassigned to the Coast Division, where they were used in a dual-purpose role. Although the test lasted only a couple of months, it was a preview of what was to come.

By the summer of 1956, all 16 H24-66s were reassigned to the Coast Division, where they began to supplement 4-6-2, 4-8-2, and 4-8-4 steam engines in commuter service. By late January

The Garys collection

It's Saturday morning and 4803 gets a bath at Bayshore in July 1958 – a common practice with the units serving in high-visibility commuter service.

John D. Hahn, Jr. collection

This late afternoon portrait of 4803 shows the massive tri-mount truck springs that helped to equalize the 42-inch wheelsets on uneven trackage.

1957, the last steam engine was removed from this work, and the 47-mile Bay Area commuter assignment was handled exclusively by the Train Masters and several steam-generator-equipped GP9s and SD9s. A 1956 schedule indicated 19 inbound trains to San Francisco each weekday morning. Even

Jim Shaw

Moving the U.S. mail was an important part of the San Francisco-San Jose runs and in this 1958 view at Bayshore, Calif., 4805 powers inbound Train 147, consisting of six head-end cars and a single gallery car.

TM 4809 rolls off the Mission Bay roundhouse turntable onto the ready tracks on January 4, 1960. While FM incorporated the large oscillating safety light into the short hood, the limited space available on the nose of GP9 5603 necessitated that the light be mounted on top of the hood. Both units feature an SP-designed plow.

supplemented by GP9s and SD9s, the 16 TMs were kept in nearly constant motion. Weekends were just as busy for the TMs, as they worked freight assignments in the Bay area. But by Monday morning, they were back in San Jose for the morning commuter rush.

The first major change in appearance for the Train Masters occurred in early 1960, when the units exchanged their black widow paint scheme for a combination of scarlet and gray, which SP had introduced in the fall of 1958. A February 27, 1959, photo at Mission Bay roundhouse shows 4813 in the scarlet and gray color scheme, making it possibly the first unit to be repainted. Between 1960 and 1962, the entire TM fleet was repainted, including 4811, which in June 1958 had gotten an experimental scheme of solid black with orange ends.

In 1965, as part of a systemwide renumbering plan, the Train Masters exchanged their 4800-series numbers for a 3020-3035 series, with actual renumbering taking place in November and December of that year. In the fall of 1971, 21-inch-high "SP" initials were added to the front of the short hood.

The end of the Train Master era on Southern Pacific began in December 1973 when two Amtrak-leased SDP45s,

After the smoke emitted from 4-6-2s, 4-8-2s, and 4-8-4s, the cloud of dust stirred up by H24-66 4812 hustling commuters home is tame in comparison. The high-mounted train indicator boards were common to units 4810-4815. Brisbane, Calif., 1958.

Dave Lustig/George Melvin collection

above, *Although their assignment remained the same, the 1960s brought a change of color to the Train Masters as SP adopted a gray and scarlet paint scheme. H24-66 4806 sits at Taylor Yard in Los Angeles on May 23, 1965.*

Waiting to depart with Train 128, *4811 carries its road number prominently displayed on the end of the short hood. San Francisco, August 26, 1964. The massive front pilot of the TM must have given crew members a sense of security in the event of a grade crossing accident.*

below, *Now wearing its third road number – TM-3, SP 4800, and SP 3020 – the former Fairbanks-Morse demonstrator retains its distinctive open walkway in this view at San Francisco, Calif., in July 1967.*

Insert photo, Dave Engman/The Garys collection, bottom photo, The Garys collection

Wearing new "SP" nose lettering, 3020 awaits the afternoon commuter rush at San Francisco on April 5, 1973, in its last year of operation. The locomotive was sold for scrap to Chrome Crankshaft in August 1974 and cut up in the following month.

Gordon Lloyd, Jr.

Gordon Lloyd, Jr.

SP 3023 is at San Jose, Calif., on April 14, 1973.

part of a fleet of 10 that SP had acquired for intercity service before Amtrak, were returned. SP then placed them into the San Francisco commuter pool. These were followed by seven more SDP45s, also returned from Amtrak, between January and April 1974. In addition, SP placed an order with EMD for a trio of GP40P-2 units to replace the TMs. The arrival of these units allowed SP to retire four Train Masters in December 1973-January 1974; and five more in June 1974. Between June and December, three more units were retired, leaving only 3021, 3022, 3031, and 3035 active. By January 1, 1975, only 3031 remained in service. On February 5, 1975, the last operating SP TM returned to San Francisco with Train *147* and was shut down. Although this ended the TM era on SP, the last H12-44 was retired on May 6, 1975, outlasting its larger Fairbanks-Morse cousin by three months.

Although their opposed-piston engines were silent, the Train Masters found another use – almost. In December 1974, 3027 was outshopped by SP's West Colton shops as MW9100, a braking sled. Mated to an Alco C-630, the unit was used in braking long cuts of cars in the West Colton hump bowl. Plans called for four sleds to be constructed, but the program was canceled before 3034, 3025, and 3033 were rebuilt to MW9101-9103. The unfinished units were subsequently scrapped at Sacramento in 1978-79. ★

The Garys collection

During their long careers on SP, the carrier made several modifications to the Train Master fuel tanks, including the use of various types of gauges. Bayshore, Calif., 1970.

Scheduled to depart San Francisco at 10 pm, Train 150 has drawn H24-66 3033 for power. The big FM's 2,400 horsepower will have little trouble in making the 1 hour, 20 minute scheduled run to San Jose this evening.

All photos this page, Gordon Lloyd, Jr.

On the turntable leads at Bayshore Yard on March 11, 1973, are Train Masters 3032 and 3034. Note the varying heights of the train indicator boards.

Variety – a line of Train Masters headed by 3034 is flanked by Alco RS-32 4000 and EMD GP9 3551 at San Jose, Calif., on July 29, 1972.

Amid a cloud of exhaust and silicon dust, Virginian 55 drills a cut of coal hoppers at Elmore, W. Va., in April 1956. For nearly 20 years, most of the carrier's 25 TMs called the mountains of West Virginia home, even though they changed owners, paint schemes and road numbers.

Virginian

Number of Units: 25

Road Numbers: 50-74

Years Built: 1954, 1957

The Virginian Railway joined the Pittsburgh & West Virginia and Akron, Canton & Youngstown as one of only three railroads to choose Fairbanks-Morse as the primary builder with which to dieselize its locomotive roster. But unlike AC&Y and P&WV, which chose FM during the early days of diesel design when all builders were considered more-or-less equal, Virginian did so after the initial tide of dieselization had passed, and the merits of one builder over another were well documented and recognized by the industry. So why did the West Virginia coal-hauler choose Fairbanks-Morse?

The 600-plus-mile Virginian consisted of two operating divisions: the flat-profiled Norfolk Division, with a westbound ruling grade of 0.6 percent, and the mountainous New River Division, with grades reaching 2.07 percent. While many of the grades of the New River Division were tamed with electric locomotives (the 134-mile stretch between Roanoke, Va., and Mullens, W. Va., was electrified in 1925), Virginian still relied on steam locomotives – primarily a fleet of 42 USRA-designed American Locomotive Co. 2-8-8-2 Mallets – to move coal from points north and west of Mullens on the New River Division. These lines, both the main line to Deepwater, W. Va., where the road connected with Chesapeake & Ohio and New York Central, and branch lines, featured grades commonly in excess of 2 percent.

On the ready tracks at Mullens, W. Va., on September 7, 1958, are H24-66s 70 and 68; 113, an Alco/Westinghouse EL-1A model built in March 1926; and 136, a General Electric EL-C built in December 1956.

Representative of Virginian's first H24-66 order, 51 displays the as-built appearance of these Phase Ib units. Note the "DE-51" on the small stainless steel plates mounted on the corners of the unit; at first they appeared only on the front corners. Tralee, W. Va., March 14, 1956.

By the early 1950s, the cost to maintain the U-class Mallet fleet was soaring. With the newest group of Mallets, the 15-member USB class, having been built in 1923 and the balance of the fleet built in 1919, it was becoming increasingly difficult to purchase parts to keep these locomotives in service. Coupled with the fact that steam locomotive maintenance and operation were labor-intensive and post-war labor costs were rising sharply, Virginian could no longer afford to fuel its locomotive fleet with cheap on-line coal, even though coal accounted for more than 95 percent of the carrier's revenues in the early 1950s.

When Virginian executives went to the Railway Supply Manufacturers' Association convention at Atlantic City, N.J., in June 1953, their intent was to meet with the Fairbanks-Morse group to discuss purchasing high-horsepower locomotives to dieselize the road's non-electric lines and coal branches. The carrier was looking for a single-unit locomotive with a rating of more than 2,000 horsepower. With Virginian's sharp curves, the unit had to be compact – the Train Master was just what the railway was looking for.

After many months of negotiations and a demonstration tour by TM-1 and TM-2, Virginian's board of directors in December approved a $5,305,931 expenditure to buy 19 Train Masters and six H-16-44s – it was FM's second largest single order. The TMs would be assigned

to the west end of the system, while the six 1,600-horsepower road switchers would replace aging M-class Baldwin 2-8-2 Mikados on the Norfolk Division.

Built between March and May 1954, the units in Virginian's first Train Master order carried road numbers 50-68, and they were classified DE-RS (Diesel Electric, Road Switcher). Virginian specified General Electric rotating equipment to match the gear used on its recently delivered (1948) 6,800-horsepower GE EL-2B streamlined electrics. These were the first Train Masters to be built with GE equipment, and the Model 752 traction motors were mounted on 40-inch wheelsets.

The Phase Ib carbodies wore Virginian's black-and-yellow color scheme. The red and white corporate herald was situated on each end of the

The quietness of the diesel-electric when approaching grade crossings, compared to the Mallets they replaced, may have accounted for the Train Masters' colorful paint scheme and large safety stripes on the pilot. H24-66s 52 and 53 rumble by the Mullens, W. Va., train station in April 1956.

left, *To fuel the new diesels, Virginian installed servicing facilities at Mullens, W. Va., in early 1954. In service only a couple of months, a grime-coated 51 takes on fuel at Mullens on May 18, 1954. The following year, new sand handling and drying facilites would be installed here.*

William A. Swartz/M. D. McCarter collection

below, *Although the four-axle FMs were generally assigned on the east end of the system, H16-44 31 was mated to TM 55 at Mullens, W. Va., in April 1956.*

Jim Shaw

carbody under the headlight. "Virginian" was spelled out on each side in 24-inch-high imitation gold lettering on a black band on the basically yellow carbody. Curiously, the road number did not appear on the carbody, only in the lighted numberboards – the same as with the EL-2B electrics.

With the arrival of the Train Master, Virginian retired 35 U-class Mallets. While the TMs were rated at only 2,400 horsepower as compared to the Mallet's 4,500-horsepower rating, each of the 19 TMs generated 98,625 pounds of tractive effort, which compared favorably to the Mallet's compound rating of 101,300 pounds. With their high availability for service, fewer diesels were required. On the eastern end of the system, the H16-44s also left their impact on steam operations. The six H16s replaced nine M-class Baldwin 2-8-2s.

In 1955, large-scale steam operations came to an end as 24 H16-44s replaced 16 M-class 2-8-2s, eight massive and modern (1945) AG-class 2-6-6-6s, and the five remaining U-class Mallets. Also set aside were the carrier's newest steamers, five BA-class 2-8-4 Berkshires. These locomotives and the 2-6-6-6s remained on the roster until 1960, when they were scrapped. A traffic upsurge in 1956 saw a handful of steamers once again under fire, but eight additional H16-44s delivered in November and December 1956 all but ended the steam era on the Virginian Railway. The last engine under steam was 0-8-0 switcher 251, which was retired in June 1957.

The upswing in export coal tonnage that started in 1956 continued into 1957. As a result, Virginian placed orders with FM for six additional Train Masters and a pair of wreck-replacement H16-44s for 1957 delivery. Twelve EL-C 3,300-horsepower C-C road-switcher electrics were also ordered from General Electric.

Built in May and June 1957, these TMs featured Phase II carbodies and were numbered 69-73, after the first order. Painted the same as 50-68, they were assigned to the same DE-RS class.

This front view of 60 shows the as-built numberboard configuration. The clips were replaced by rubber gaskets by 1956. Mullens, W. Va., May 18, 1954. Note the poling pockets cast into number 60's pilot; the practice of poling was later eliminated because of its potential for danger.

William A. Swartz/M. D. McCarter collection

John D. Hahn, Jr.

above, *Bringing a loaded coal train down the Guyandot River Branch, number 65 will turn its train over to an electric locomotive for the continued trip eastward to Roanoke. Ultimately, the train will tie up at the docks at Sewalls Point near Norfolk, Va. Elmore, W. Va., March 14, 1956.*

right, *Relaying empty cars to the mines, 67 pauses at Mullens, W. Va., in April 1956, for a fresh crew.*

Jim Shaw

When Norfolk & Western absorbed Virginian, only a new road number and the application of that number below the cab window denoted the change in ownership for the Train Masters. Mullens, W. Va., June 12, 1960.

These would be the last new C-C truck locomotives purchased by the railway, as merger talks with Norfolk & Western Railway, begun in 1925, finally came to completion on December 1, 1959.

To avoid number conflicts with N&W Alco T-6 switchers, the 25 Virginian Train Masters were renumbered 150-174. Their assignments remained much the same after the merger, with all of the units working out of Mullens, W. Va. For awhile, the units remained painted yellow and black with only the 150-series road numbers denoting a change. Eventually, they were repainted into N&W's solid black, and later solid blue,

left, *Renumbered electric and diesel-electric locomotives pose outside the Mullens shops on October 14, 1960.* below, *H24-66s 169 and 154 are at Mullens, W. Va., on June 12, 1960.*

R. D. Patton/The Garys

but they continued to work the West Virginia coal fields.

In 1967, five Train Masters were traded to Alco for a like number of C-630 road freight units. Included were Virginian/N&W 151, 157, 160 and 166 (the fifth unit was Wabash 3599, which had swapped numbers with 161). The big Centuries came from Schenectady riding on tri-mount trucks from the retired Train Masters. The experiment was never repeated, as the remaining TMs continued working, with seven traded to EMD on new SD45s in 1969 and 1970. The last operating unit, 173, was set aside on July 5, 1976, the day after operating on a fan trip out of Roanoke, Va., mated to Bicentennial-painted SD45 1776.

Toward the end of their careers, several of the former Virginian and Wabash (also merged into N&W) Train Masters swapped road numbers to meet equipment trust agreements as units whose trusts had not yet expired were traded in on new locomotive orders. But N&W found a new use for the Train Masters – slug conversion.

Between 1971 and 1981, N&W rebuilt 14 former Virginian and six Wabash Train Masters into RP-F6 slugs at its Roanoke (Va.) shop. The frame, trucks, and traction motors of the original TM were retained, while a

Gordon B. Mott/Louis A. Marre collection

Gordon B. Mott/Louis A. Marre collection

top, *Former Virginian 62 wears the paint and lettering scheme of its new owner at Mullens, W. Va., on November 7, 1961.*

middle photos, *A problem common to aging Train Masters was their propensity to start brush fires. To reduce this problem, N&W fitted screen-type spark arrestors to the twin exhaust stacks. Units 165 and 170 are at Mullens, W. Va., on April 6, 1968.*

bottom, *N&W 158 sports a larger, more effective spark arrestor design at Mullens, W. Va., on March 24, 1975. The unit, retired in June 1976, emerged from Roanoke Shops in September of that year as a RP-F6 slug.*

Gordon Lloyd, Jr.

Virginian Train Master Renumberings

VGN NUMBER	FIRST N&W NUMBER	SECOND N&W NUMBER	THIRD N&W NUMBER
50-52	150-152		
53	153	3598:3	
54-60	154-160		
61	161	3599:2	
62-64	162-164		
65	165	169:2	
66	166		
67	167	3598:2	156:2
68	168		
69	169	165:2	
70-74	170-174		

new, lower profile carbody was fabricated. The slugs were mated with specially modified EMD SD40s and Alco C-630s for use in flat switching and as hump sets. Renumbered 9900-9919, the units worked at various yards across the N&W system.

The RP-F6s retained their 9900-series numbers with the formation of Norfolk Southern. By the 1990s, their original FM-designed tri-mount trucks were wearing out and consideration was given to fitting the units with six-axle EMD trucks. The cost of such a conversion was prohibitively high and by early 1997, only three units (9902, 9909, and 9910) remained in service. One unit, 9914, was donated to the Virginia Transportation Museum in Roanoke. ★

Both photos, Gordon Lloyd, Jr.

above, **One Norfolk & Western Train Master lasted long enough to be repainted in the carrier's contemporary "NW" scheme – 173 continues to work at Mullens, W. Va., on September 5, 1974, 17 years after being built by Fairbanks-Morse.** insert, **Unit 173 is at Mullens, W. Va., on March 24, 1975.**

James C. Herold/Louis A. Marre collection

left, **Although never equipped for passenger service, the sole surviving Virginian Train Master had the honor of pulling a special passenger excursion on July 4, 1976. The train was stopped at Stoneville, Va.**

Former Fairbanks-Morse Train Master demonstrator TM-1 poses at Decatur, Ill., on February 14, 1956, wearing the colors and road number of its new owner, Wabash. In a few months, 550 and sister unit 551 (formerly TM-2) would be joined by six new Train Masters, as Wabash found the unit to be well-suited for its operations. Note the blanked-out dynamic brake equipment openings.

J. R. Quinn/courtesy of Overland Models

Wabash

Number of Units: 8

Road Numbers: 550-554, 552A-554A

Years Built: 1953, 1956

The eight Train Masters owned by Wabash Railroad stand in stark contrast to the rest of that company's diesel fleet. They rode on C-C trucks, the only Wabash units so equipped, and they were the heaviest diesels on the property, weighing more than 378,000 pounds. Their place on the Wabash roster can probably be attributed to the Fairbanks-Morse sales department, which marketed the Train Master not only as a jack-of-all-trades locomotive, but also, with its high horsepower rating, as a means to achieve unit reduction. To back up these claims, FM arranged for a demonstration tour of TM-1 and TM-2. So successful were they that in February 1954, the pair of demonstrators became the first Train Masters purchased by the carrier.

This wasn't the carrier's first FM acquisition – a pair of H10-44s was delivered in 1946, followed by a second pair in 1949. These were followed by three H12-44s in 1953, but Wabash had relied on EMD and Alco to supply the majority of its road freight diesels and yard units, with FM, Lima, and Baldwin contributing only a handful of switchers. After the war, Wabash began dieselizing its passenger operations with EMD E7s and E8s and Alco PAs. In 1949, Wabash began dieselizing its mainline freights with A-B-A sets of EMD F7s and Alco FA/FBs, and by the end of 1953, regular steam operations had ended.

It was against this backdrop that Wabash purchased its first Train Masters. Repainted blue, white, and gray, the former demonstrators were renumbered 550 and 551. Equipped with steam generators, the TMs were versatile units that could be used as standby passenger power, used singly on secondary freights, or mated to replace an A-B-A set of 1,500-horsepower cab units in fast freight service. The dynamic braking equipment they originally carried

R. R. Wallin

Shortly before being renumbered 556, H24-66 554A leads an "extra" freight, assisted by a pair of Alco FAs and an EMD F7 at Hull, Ill., in 1961. Unlike other carriers, which demoted minority-make locomotives to less-demanding tasks, Wabash worked its TMs in mainline freight service during their entire careers.

was removed before delivery to Wabash. They must have been successful – a little more than a year later, Wabash placed an order for six more Train Masters.

Wabash was an east-west railroad whose main line connected Buffalo, N.Y., with Chicago, Kansas City, and St. Louis. Its profile was relatively flat with the exception of the line between Decatur, Ill., and Moberly, Mo., which had a ruling grade of 1.2 percent. Wabash's original idea was to use the Train Master in pairs, their 4,800 combined horsepower being slightly higher than the 4,500 horsepower rating of an A-B-A cab-unit consist, but with one fewer unit. For this role, they were assigned road numbers 552-552A, 553-553A, and 554-554A.

With the two former demonstrators on its roster, Wabash encountered first-hand the problems of mating a TM to another builder's model. Noting the way Southern used its TMs as booster units between sets of F units, Wabash specified that its order be built with low end platforms like the Southern units. The units, built with Phase II carbodies, were equipped with General Electric rotating gear. For their dual-service role, they were equipped with Vapor Corp. Model 4740 steam generators. Like the former demonstrators, they were painted in Wabash's intricate blue, white, and gray scheme. Unlike 550 and 551, which were assigned Wabash class D24 (Diesel, 2,400 horsepower) these units were assigned class D48 (Diesel, 4,800 horsepower), reflecting Wabash's plans to operate the units in pairs.

Louis A. Marre

The open space below the walkway and large oscillating safety light above the headlight are quick spotting features of the former TM demonstrators. Decatur, Ill., January 3, 1961.

Louis A. Marre

Featuring low end platforms, Wabash's six purchased-new Train Masters were equipped with a twin-beam safety light mounted above the headlight on the short hood only. Unit 554 was at Wabash's main shop facility at Decatur, Ill., on January 3, 1961.

The Train Masters went to work moving freight between Detroit and Kansas City, and Chicago and Kansas City, and as FM had advertised, they were capable of moving hotshots at speed or working wayfreights with numerous switching moves enroute. Among the passenger assignments handled by the TMs was Train *112/113*, an unnamed all-stops overnight local between Decatur and Chicago. In early 1961, the pairs were renumbered 552-557 as they were used in mixed lashups with other models rather than exclusively in pairs, as originally planned.

The next change to the TMs came in early 1964, when they were shipped to Alco's Schenectady, N.Y., plant to be repowered with 16-cylinder Alco 251B engines. Retaining their original Westinghouse/General Electric electrical gear (although some reports indicate that 551 was returned with a GE main generator), the units were returned to Wabash between March and May, now rerated at 2,350 horsepower and wearing a new paint scheme and road numbers. Painted solid blue, they were renumbered 592-599 (598 and 599 were former FM demonstrators TM-1 and TM-2) to make room for numbers assigned to new GP35s purchased by the carrier. The units returned to their original assignments, moving mainline freight, but this was about to change as Wabash became part of Norfolk & Western Railway system on October 16, 1964.

On N&W, the Train Masters were

Louis A. Marre

This rear view of 553 at Decatur, Ill., on January 3, 1961, shows the single headlight and end platform arrangement with the drop step and MU receptacle arrangement. To match the two demonstrators, Wabash specified 42-inch wheel sets on GE-equipped TMs.

Former Wabash 557 poses outside Alco's Schenectady, N.Y., facility in April 1964 after its opposed-piston engine was replaced by an Alco 16-cylinder 251-series prime mover. The only external changes made to the H24-66 were to the engine compartment hood, where the original louvers were replaced by screened openings with the filter media clearly visible.

Alco

assigned road numbers 3592-3599. While most of the units were renumbered in 1965, 599 was out of service and was never renumbered. With new second-generation road locomotives being delivered yearly, the hybrid TMs were reassigned to the recently expanded Bellevue, Ohio, hump yard. This former Nickel Plate facility took on fresh importance with the merger of Nickel Plate, Wabash, and Norfolk & Western.

The next change to the Wabash fleet occurred in 1967 when 599, out of service since late 1966, swapped its power plant and engine compartment hood with former Virginian H24-66 61, now N&W 161 – 599 was renumbered 161 and traded to Alco on a Century 630 order, and 161 was renumbered 3599. In 1970, 3598 swapped numbers with 167, with the Alco-repowered unit traded to EMD on an SD45 order. Other number swaps took place between the Wabash units (see table) as N&W tried to keep the TMs in a solid number block.

The six remaining units continued to work at Bellevue until the mid-1970s, when they were removed from service

R. R. Wallin/Louis A. Marre collection

After rebuilding, the repowered Train Masters remained on mainline freight service, but their days were numbered as N&W concentrated its hybrid units in yard service at Bellevue, Ohio. Decatur, Ill., July 1964.

right, *Between assignments, N&W 3597 and 3594 are at the Bellevue, Ohio, engine terminal on May 5, 1968. Wabash's repowered TMs spent nearly their entire N&W careers working the Bellevue hump and even after retirement, a few returned as 9900-series slugs.*

Jim Wozniczka

With an Alco 251-series engine beating inside its carbody, 594 rests outside Decatur (Ill.) shops on March 9, 1965. Wabash was the only railroad to repower its Train Masters.

Louis A. Marre

Former Wabash 554 serves as a bumping post at N&W's Shaffers Crossing facility on November 20, 1977. Next stop for this TM will be the carrier's Roanoke Shops for conversion into a yard slug.

Robert F. Graham

At first glance, N&W 3598 could be identified as a former Wabash Train Master judging from its 3500-series road number, but closer inspection reveals that it is a former Virginian TM (either 53 or 67) with its opposed-piston power plant intact. Mullins, W. Va., August 9, 1970.

John D. Hahn, Jr.

Wabash Train Master Renumberings

FIRST WABASH NUMBER	SECOND WABASH NUMBER	FIRST N&W NUMBER	SECOND N&W NUMBER
550	598	3598	167:2
551	599	3599	161:2
552	592	3592	3598:4
552A	595	3595	3597:2
553	593	3593	
553A	596	3596	
554	594	3594	
554A	597	3597	3595:2

and shipped to N&W's Roanoke, Va., shop to be converted into slugs. Between 1974 and 1981, they emerged as RP-F6 slugs riding on their original tri-mount trucks, but with new carbodies and 9900-series road numbers. Retirements of the ex-Wabash Train Master slugs began in 1993, with one unit, 9910 (ex-Wabash 553A) remaining active at Bellevue into early 1997. ★

At first glance, Chicago & North Western 1678 looks like a Train Master, but close examination of the rear carbody section reveals a missing rooftop fan and a blanked-out radiator section. Less noticeable is the H16-66's shorter length when compared to the H24-66 – 62 feet versus 66 feet.

H16-66

Fairbanks-Morse introduced a line of 1,500-horsepower road switchers in 1947 with its newly developed eight-cylinder opposed-piston engine. But the road-switcher carbody had yet to develop a niche on most carriers' locomotive rosters – between 1947 and 1949, FM sold only 35 of these units to nine railroads.

Starting in 1950, the output of the eight-cylinder OP engine was raised to 1,600 horsepower, and the builder accordingly introduced the H16-44 model. The H16-44s proved to be successful – 299 of them were built over a 13-year span, making it the second most popular model in the FM line (only the H12-44 switcher outsold the H16-44, with 336 units built between 1950 and 1961).

At the same time as FM was introducing the H16-44, Chicago & North Western approached the company about building a six-axle version of the H16-44 to limit the axle loading to 50,000 pounds per axle. C&NW's relationship with on-line locomotive builder Fairbanks-Morse started in November 1944 when FM delivered its second locomotive, an H10-44 assigned C&NW number 1036. Four more 1,000-horsepower switchers were delivered to North Western in 1946, followed by four Erie-built passenger cab units in 1947. By 1950, 26 1,000-horsepower and two 1,200-horsepower switchers worked in yards throughout the system. With a host of branch lines requiring a lightweight locomotive capable of bi-directional operation, North Western became interested in the then-new road-switcher models being offered by Alco, Baldwin and EMD. C&NW also talked to FM about customizing its H16-44 with six-axle trucks to reduce axle loadings. By equipping the unit with General Steel Castings' rigid-frame six-wheel truck, FM was able to meet a customer's needs with minimal engineering effort.

The result was FM contract number LD-93 for five H16-66s, road numbers 1510-1514, for C&NW and a single unit, numbered 150, for a C&NW subsidiary, Chicago, St. Paul, Minneapolis & Omaha. Delivered in January and February 1951, these units featured Phase Ia carbodies with industrial designer Raymond Loewy's distinctive styling touches – curved cab windows, slightly sloped ends, and raised headlight mounts. Interestingly, the first two units, 1510-1511, had builders numbers 16-L-35 and 16-L-37 from FM's old number series, while the balance of the order carried 16-L-275 through 16-L-278 from the new series. The use of Westinghouse

The first H16-66s were basically H16-44s on General Steel Castings C-C trucks (the same design used on many Baldwin C-C units). C&NW 1511, at Ironwood, Mich., on April 19, 1965, displays the Phase Ia carbody with its distinctive rounded cab windows.

While the basic H-series carbody design was retained on Chicago & North Western's second H-16-66 order, the cab windows were simplified to reduce construction costs. C&NW 1609, at Ironwood, Mich., in June 1960, is typical of the Phase Ib H16-66.

electrical equipment required that 42-inch-diameter wheels be installed to accommodate the Westinghouse Model 370 traction motors. Weighing 287,000 pounds, these locomotives were equal in weight to Baldwin's comparable model, the DRS-6-6-1500, but slightly heavier than Alco's 278,900-pound RSD-4, delivered at the same time.

Satisfied with the performance of the first six units, C&NW ordered eight more in 1952 under FM contract number LD-137. These units, numbered 1605-1612, arrived from Beloit in October and November. They were built with Phase Ib carbodies – the rounded cab window design was replaced with a more economical square style.

In 1953, C&NW placed a third H16-66 order. Eleven units, built on contract numbers LD-150 and LD-151, arrived in July, six of them, numbers 1668-1673, for North Western, and five, numbers 168-172, for its "Omaha Road" subsidiary, CStPM&O. These were not only the last H16-66s to ride on GSC trucks, but they also were the last units built with the basic H-series carbody introduced in 1944.

With the coming of the Train Master in 1953, the H16-66 was offered with a shortened version of the Train Master carbody in August 1953. Four feet shorter than the H24-66, the new H16-66 quickly acquired the nickname Junior Train Master or Baby Train Master among

C&NW assigned six H16-66s to its Chicago, St. Paul, Minneapolis & Omaha subsidiary, including 171, one of 11 units delivered in mid-1953. Other than number series and lettering, they were identical to their C&NW counterparts.

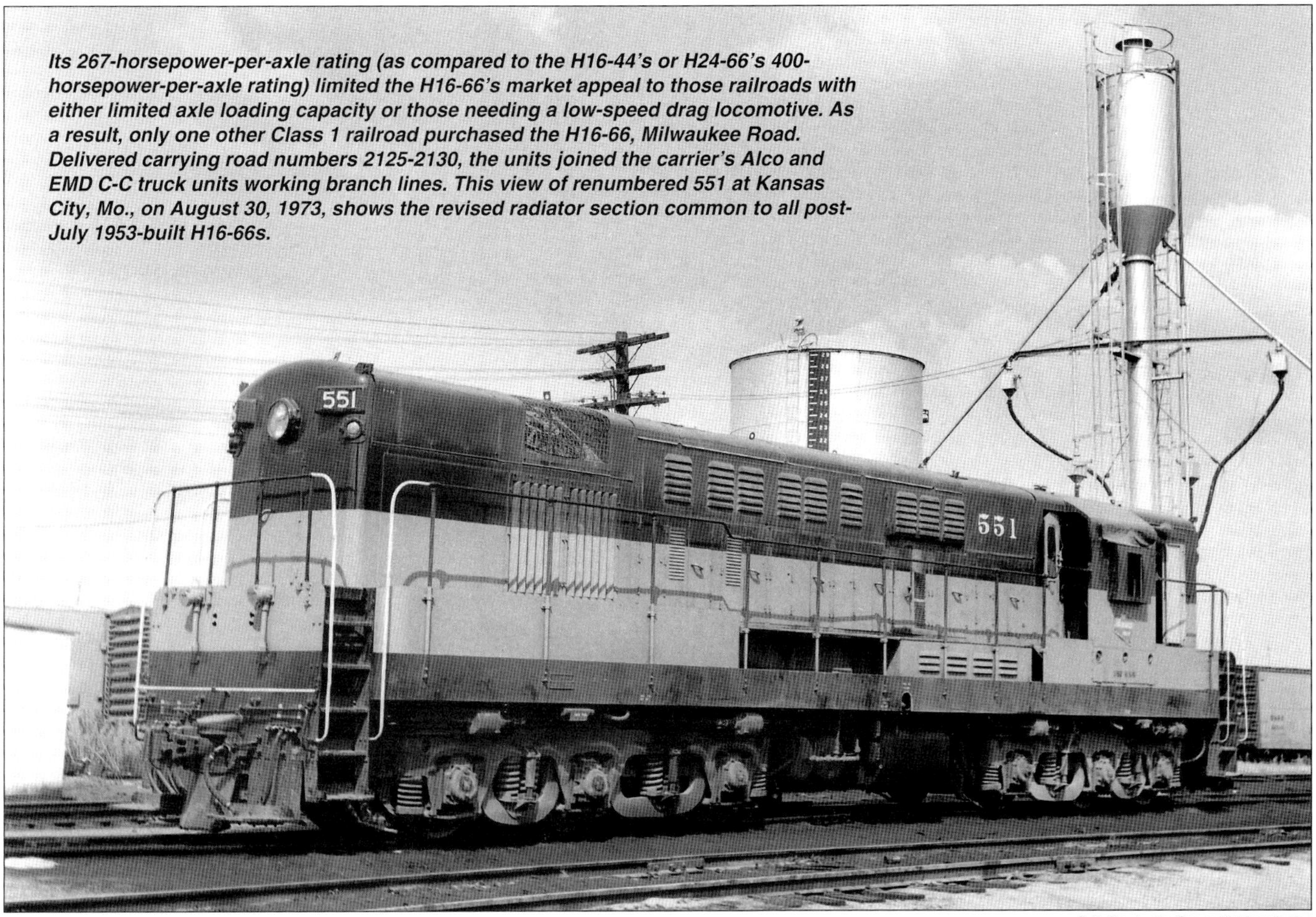

Its 267-horsepower-per-axle rating (as compared to the H16-44's or H24-66's 400-horsepower-per-axle rating) limited the H16-66's market appeal to those railroads with either limited axle loading capacity or those needing a low-speed drag locomotive. As a result, only one other Class 1 railroad purchased the H16-66, Milwaukee Road. Delivered carrying road numbers 2125-2130, the units joined the carrier's Alco and EMD C-C truck units working branch lines. This view of renumbered 551 at Kansas City, Mo., on August 30, 1973, shows the revised radiator section common to all post-July 1953-built H16-66s.

C. N. Shankweiler/George Melvin collection

railfans. FM referred to the unit in its advertising brochures as a "1,600-horsepower General Service" model. Although shorter than the Train Master, the redesigned H16-66 was 62 feet long as compared to the 55-foot, 8-inch length of the earlier Loewy-styled units. The H16-66's eight-cylinder diesel engine required less cooling capacity and so only three cooling fans, instead of four, were enclosed in the down-sized radiator section. Like the Train Master, the H16-66 featured stepped handrails along the long hood, but unlike the design of the production Train Masters, open space appeared below the walkways.

The unit was available in a wide variety of weights, from 298,000 pounds to 375,000 pounds. The standard fuel tank featured a capacity of 1,200 gallons, with optional capacity as high as 3,600 gallons. For passenger service, a 4,500-pound/hour steam generator was available with a 2,400-gallon boiler water supply tank. Three gear ratios were offered: 62:17, 63:15, or 68:15. While 40-inch-diameter wheels were standard on the GE-equipped units, 42-inch wheels were available as an option.

Milwaukee Road became the first customer for the redesigned H16-66, ordering six units for delivery in August and September 1953 under contract number LD-153. In addition to the new Phase IIa carbody style, FM adopted its TM-developed tri-mount truck as standard on the new H16-66. Numbered 2125-2130, the units were equipped with Westinghouse electrical equipment, the last H16-66s built with this gear, as FM made the switch to General Electric equipment. The Westinghouse equipment included a Model 497BZ main generator (the same as used on Milwaukee's C-Liners) and six Model 370 traction motors mounted on axles with 42-inch wheels. Intended for branchline service, the units were built with small 800-gallon fuel tanks and weighed 298,000 pounds. Equipped with multiple-unit connections, the H16-66s could and did MU with the FM

Dennis Schmidt

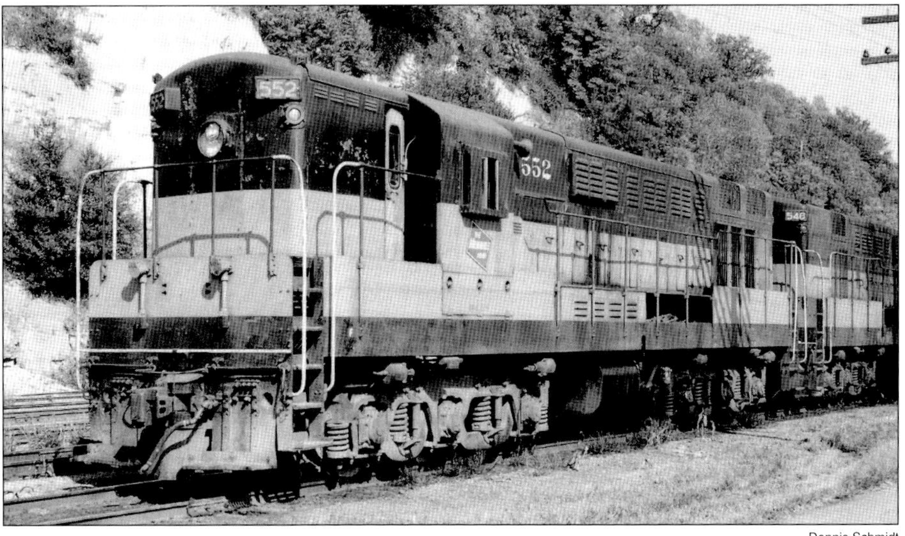

Although dropped on Train Master production units, the open space below the walkway seen on Milwaukee Road 552 at St. Paul, Minn., on September 2, 1971, was found on all Phase II H16-66s.

Chicago & North Western's first order for the redesigned H16-66 model began arriving in July 1954. Unlike Milwaukee's units, these came equipped with dynamic brakes and full-size fuel tanks. As built, the tri-mount trucks lacked flanged sideframes, but C&NW elected to retrofit its H16-66s with this design. Ishpeming, Mich., October 5, 1968.

Unlike 1674 above, C&NW had yet to enclose the space beneath the walkway of 1675 at Iron Mountain, Mich., on June 17, 1970. The numberboards on these Phase IIa H16-66s were attached with clips, while the Phase IIb numberboards were held in place with rubber gaskets. The letter "B" between the 16 and 75 denotes C&NW's unit weight classification for certain line restrictions. The B designation means that the unit weighs between 300,000 and 329,000 pounds – C&NW listed the actual weight of these H16-66s as 298,000 pounds.

C-Liners that Milwaukee had purchased a few years earlier, although the H16s did not have dynamic brakes. Painted orange and brown, the units displayed the Milwaukee Road shield under the cab side windows and on both ends.

The following year, Chicago & North Western returned for another batch of H16-66s. Ten units, numbered 1674-1683, were delivered in July and August 1954 on contract LD-169. Built with split radiator cooling fans (the Phase IIa carbody), these units were equipped with GE electrical equipment and dynamic brakes. At 298,000 pounds, these new-style H16-66s were heavier than the earlier units.

Although the H16-66 found no new buyers, C&NW continued to order the model for the next two years. The 1955 order on contract LD-182 consisted of 10 units numbered 1691-1700. Delivered between August and November, units 1691-1695 were equipped with dynamic brakes, while 1691-1693 and 1700 came with steam generators for commuter service. Like previous H16-66s built for North Western, 1696-1699 were lightweight units, lacking dynamic brakes and steam generators, for branchline operation. As the Train Master carbody evolved during its production run, so to did the H16-66 carbody. On this order, the radiator cooling fans were built with closer spacing. This is the Phase IIb H16-66 carbody. On some

R. C. Anderson/*Diesel Era* collection

Only four H16-66s, including 1692 at Iron Mountain, Mich., on March 16, 1970, were equipped with steam generators for commuter/passenger service. Unlike the Train Master whose boiler water fill spout was centered in the short hood, the H16-66's water fill was offset to the engineer's side.

units, C&NW modified the engine compartment ventilation louvers.

C&NW's last order was for 12 units, but with dieselization nearly complete and the carrier implementing an improved locomotive utilization plan, the quantity was reduced to six units. Built on FM contract LD-197 and numbered 1901-1906, these were identical to the previous Phase IIb order with closely spaced radiator fans. The open space beneath the long-hood walkway, long since discontinued on Train Master production, was continued on these June 1956-built units.

The last chapter in the story of the H16-66 didn't start until 1958. Alcoa (Aluminum Co. of America) purchased one H16-66 unit in January 1958. Lettered for Squaw Creek Coal Co., it ran from a mine at Boonville, Ind., to Yankeetown, Ind., on the Ohio River. Running over trackage of the Yankeetown Dock Railroad, the unit was numbered 721001 on the side of the cab, with the digits 001 in the numberboards. It was not equipped with dynamic brakes or MU equipment.

The last H16-66 built was shipped 10 months later to the Tennessee Valley Authority for use at its Gallatin, Tenn., steam plant. Painted dark olive drab, it was lettered US-TVA and carried the number 24. TVA 24 switched coal hoppers that were delivered by the Louis-

right, *Sporting the carrier's experimental solid green paint scheme, 1698 displays a C&NW-modified engine louver arrangement at Ironwood, Mich., on August 21, 1962.*

William Raia/George Melvin collection

C&NW 1696 was teamed with GP7 1596 on a mixed freight at Fond du Lac, Wis., on May 29, 1971. Built on the same order as 1692, above, this unit lacks dynamic brakes and a steam generator. The short hood was designated as the front on C&NW's Phase II H16-66s, while the long hood was designated as front on the Phase I H16s.

R. C. Anderson/*Diesel Era* collection

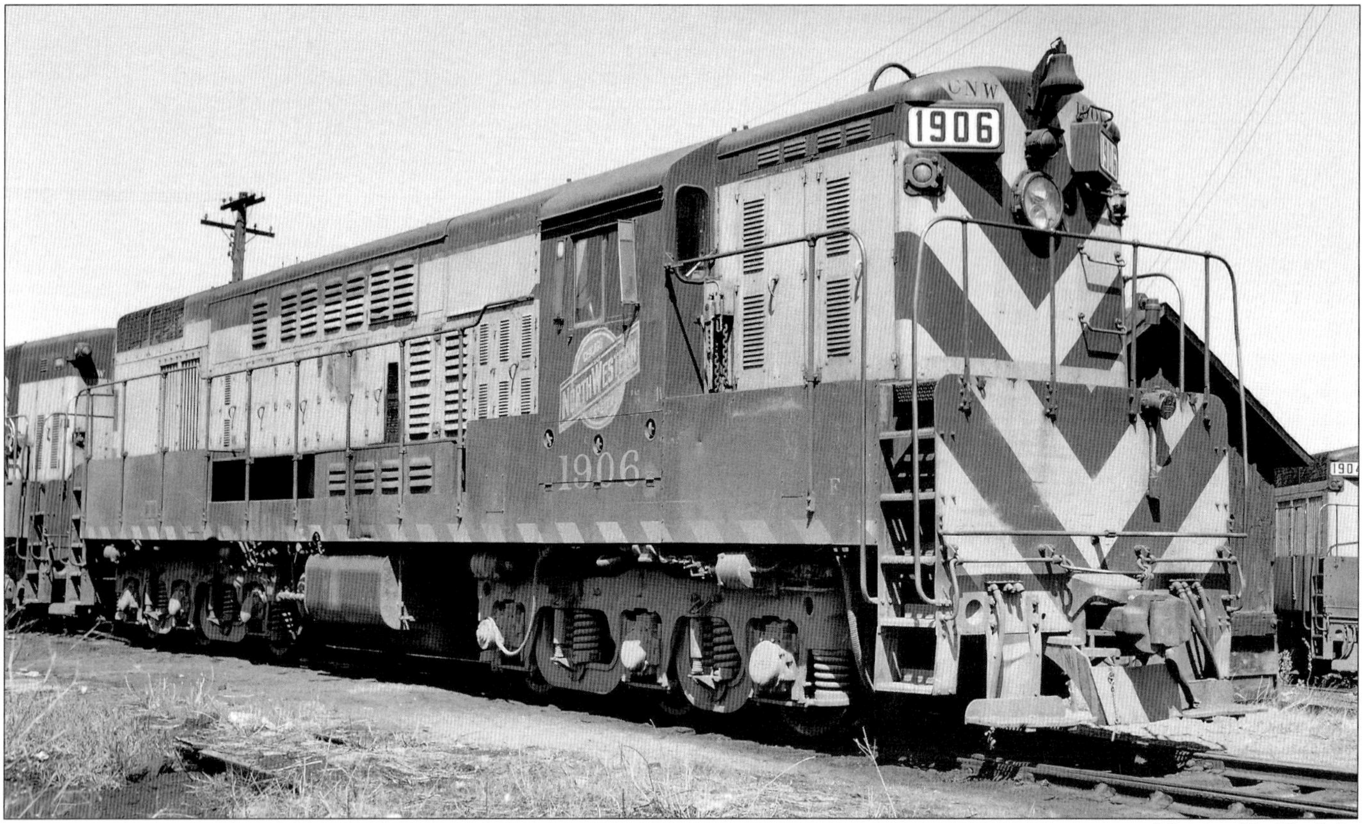

R. C. Anderson/*Diesel Era* collection

Although the pigments have faded, C&NW 1906 displays an early paint scheme with frame-edge striping and diagonal end striping. The unit was part of the carrier's final H16-66 order for six units. Iron Mountain, Mich., May 5, 1968.

R. C. Anderson/*Diesel Era* collection

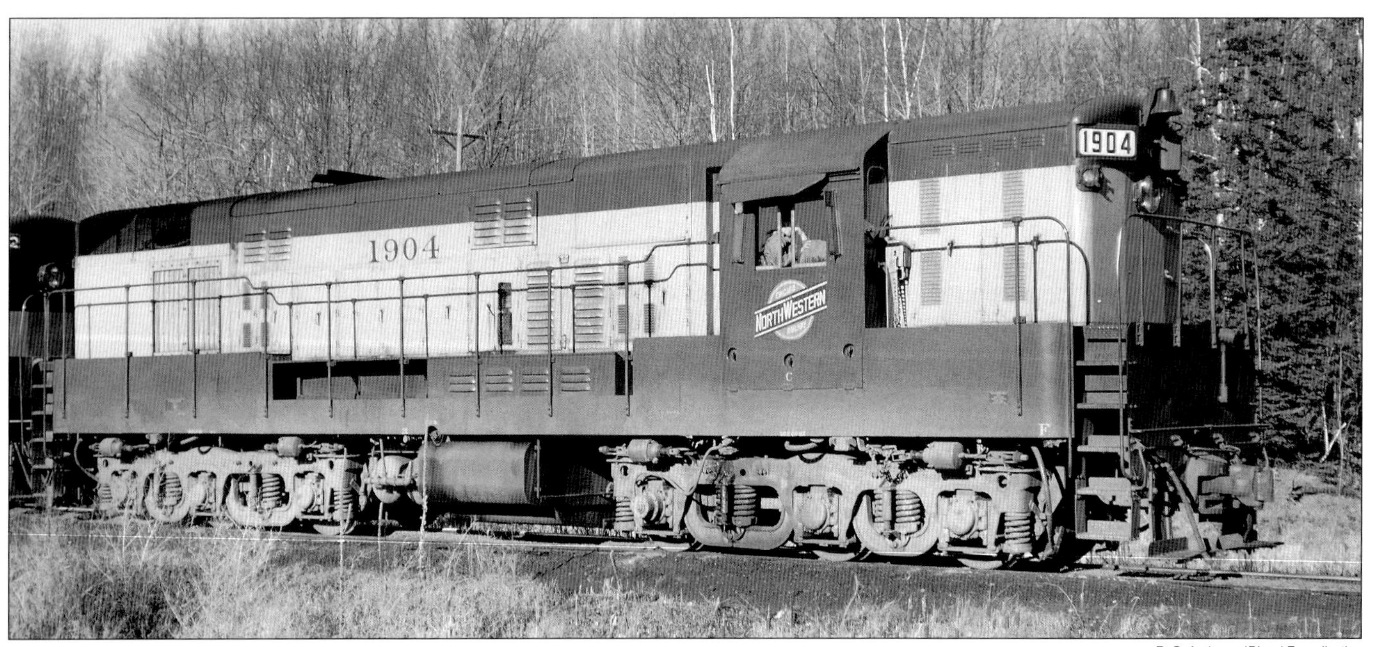

Mated to Milwaukee Road H16-66 552, C&NW 1904 is at Iron Mountain, Mich., on December 3, 1967. During their long careers on the North Western, the road's shop forces made numerous modifications to the carbodies of its FM fleet – note the removal of the engine compartment ventilation louvers across the top of the hood and directly behind the cab.

ville & Nashville Railroad to the TVA interchange, located four miles from the steam plant. Built in October 1958, TVA 24 was the last six-axle unit assembled by Fairbanks-Morse. Because this H16-66 operated alone, no multiple-unit equipment was specified.

Both industrial units were built with Phase IIb carbodies similar to C&NW's final two orders of 1955 and 1956. It has been reported that the Squaw Creek and TVA units were built from stock parts left over from the canceled portion of the 1956 C&NW order, but this theory has never been substantiated.

Of the total of 59 H16-66s built between 1951 and 1958, two are still in existence. TVA 24 is still working its original assignment at Gallatin, now wearing a gray and blue paint scheme and renumbered F3060. Painted in a Canadian Pacific scheme, Squaw Creek 721001 is preserved at the Alberta Pioneer Railway Association in Canada.

As for the Milwaukee and C&NW H16-66s, many worked for more than 20 years – longer than their 2,400-

Fairbanks-Morse H16-66 Roster

RAILROAD/ OWNER	ROAD NUMBER	2nd ROAD NUMBER	BUILDER'S NUMBER	BUILDER'S DATE	CONTRACT NUMBER	CARBODY PHASE	GEAR RATIO	WHEEL DIA.	OPTIONS	WEIGHT	FUEL CAPY.	WATER CAPY.	ELECT. EQUIP.	NOTES
Aluminum Corp. of America (Alcoa)	721001		16-L-1159	1/58	LD-209	IIb	74:18	40"			1,200		GE	
Chicago, Milwaukee, St. Paul & Pacific	2128	553	16-L-693	7/53	LD-153	IIa	68:15	42"		298,000	800		WH	1
Chicago, Milwaukee, St. Paul & Pacific	2129	554	16-L-694	7/53	LD-153	IIa	68:15	42"		298,000	800		WH	1
Chicago, Milwaukee, St. Paul & Pacific	2130	555	16-L-695	7/53	LD-153	IIa	68:15	42"		298,000	800		WH	1
Chicago, Milwaukee, St. Paul & Pacific	2125	550	16-L-757	7/53	LD-153	IIa	68:15	42"		298,000	800		WH	1
Chicago, Milwaukee, St. Paul & Pacific	2126	551	16-L-758	7/53	LD-153	IIa	68:15	42"		298,000	800		WH	1
Chicago, Milwaukee, St. Paul & Pacific	2127	552	16-L-759	7/53	LD-153	IIa	68:15	42"		298,000	800		WH	1
Chicago & North Western	1510		16-L-35	1/51	LD-93	Ia	63:15	42"		287,000	800		WH	
Chicago & North Western	1511		16-L-37	1/51	LD-93	Ia	63:15	42"		287,000	800		WH	
Chicago & North Western	1512		16-L-275	2/51	LD-93	Ia	63:15	42"		287,000	800		WH	
Chicago & North Western	1513		16-L-276	2/51	LD-93	Ia	63:15	42"		287,000	800		WH	
Chicago & North Western	1514		16-L-277	2/51	LD-93	Ia	63:15	42"		287,000	800		WH	
Chicago & North Western	1605		16-L-659	10/52	LD-137	Ib	63:15	42"		287,000	800		WH	
Chicago & North Western	1606		16-L-660	10/52	LD-137	Ib	63:15	42"		287,000	800		WH	
Chicago & North Western	1607		16-L-661	10/52	LD-137	Ib	63:15	42"		287,000	800		WH	
Chicago & North Western	1608		16-L-662	10/52	LD-137	Ib	63:15	42"		287,000	800		WH	
Chicago & North Western	1609		16-L-663	10/52	LD-137	Ib	63:15	42"		287,000	800		WH	
Chicago & North Western	1610		16-L-664	10/52	LD-137	Ib	63:15	42"		287,000	800		WH	
Chicago & North Western	1611		16-L-665	11/52	LD-137	Ib	63:15	42"		287,000	800		WH	
Chicago & North Western	1612		16-L-666	11/52	LD-137	Ib	63:15	42"		287,000	800		WH	
Chicago & North Western	1668		16-L-696	7/53	LD-151	Ib	63:15	42"		286,000	800		WH	
Chicago & North Western	1669		16-L-699	7/53	LD-151	Ib	63:15	42"		286,000	800		WH	
Chicago & North Western	1670		16-L-700	7/53	LD-151	Ib	63:15	42"		286,000	800		WH	
Chicago & North Western	1671		16-L-701	7/53	LD-151	Ib	63:15	42"		286,000	800		WH	
Chicago & North Western	1672		16-L-702	7/53	LD-151	Ib	63:15	42"		286,000	800		WH	
Chicago & North Western	1673		16-L-703	7/53	LD-151	Ib	63:15	42"		286,000	800		WH	
Chicago & North Western	1674		16-L-872	8/54	LD-169	IIa	74:18	40"	D	320,400	1,200		GE	
Chicago & North Western	1675		16-L-873	8/54	LD-169	IIa	74:18	40"	D	320,400	1,200		GE	
Chicago & North Western	1676		16-L-874	7/54	LD-169	IIa	74:18	40"	D	320,400	1,200		GE	
Chicago & North Western	1677		16-L-875	7/54	LD-169	IIa	74:18	40"	D	320,400	1,200		GE	
Chicago & North Western	1678		16-L-876	7/54	LD-169	IIa	74:18	40"	D	320,400	1,200		GE	
Chicago & North Western	1679		16-L-877	7/54	LD-169	IIa	74:18	40"	D	320,400	1,200		GE	
Chicago & North Western	1680		16-L-878	7/54	LD-169	IIa	74:18	40"	D	320,400	1,200		GE	
Chicago & North Western	1681		16-L-879	7/54	LD-169	IIa	74:18	40"	D	320,400	1,200		GE	
Chicago & North Western	1682		16-L-880	7/54	LD-169	IIa	74:18	40"	D	320,400	1,200		GE	
Chicago & North Western	1683		16-L-881	7/54	LD-169	IIa	74:18	40"	D	320,400	1,200		GE	
Chicago & North Western	1691		16-L-983	9/55	LD-182	IIb	74:18	40"	DS	324,800	1,200	1,000	GE	
Chicago & North Western	1692		16-L-984	10/55	LD-182	IIb	74:18	40"	DS	324,800	1,200	1,000	GE	
Chicago & North Western	1693		16-L-985	10/55	LD-182	IIb	74:18	40"	DS	324,800	1,200	1,000	GE	
Chicago & North Western	1694		16-L-981	11/55	LD-182	IIb	74:18	40"	D	316,000	1,200		GE	
Chicago & North Western	1695		16-L-982	11/55	LD-182	IIb	74:18	40"	D	316,000	1,200		GE	
Chicago & North Western	1696		16-L-972	8/55	LD-182	IIb	74:18	40"		295,000	1,200		GE	
Chicago & North Western	1697		16-L-973	8/55	LD-182	IIb	74:18	40"		295,000	1,200		GE	
Chicago & North Western	1698		16-L-974	9/55	LD-182	IIb	74:18	40"		295,000	1,200		GE	
Chicago & North Western	1699		16-L-987	9/55	LD-182	IIb	74:18	40"		295,000	1,200		GE	
Chicago & North Western	1700		16-L-986	9/55	LD-182	IIb	74:18	40"	S	321,300	1,200	1,000	GE	
Chicago & North Western	1901		16-L-1003	6/56	LD-197	IIb	74:18	40"		298,600	1,200		GE	
Chicago & North Western	1902		16-L-1004	6/56	LD-197	IIb	74:18	40"		298,600	1,200		GE	
Chicago & North Western	1903		16-L-1005	6/56	LD-197	IIb	74:18	40"		298,600	1,200		GE	

RAILROAD/ OWNER	ROAD NUMBER	2nd ROAD NUMBER	BUILDER'S NUMBER	BUILDER'S DATE	CONTRACT NUMBER	CARBODY PHASE	GEAR RATIO	WHEEL DIA.	OPTIONS	WEIGHT	FUEL CAPY.	WATER CAPY.	ELECT. EQUIP.	NOTES
Chicago & North Western	1904		16-L-1029	6/56	LD-197	IIb	74:18	40"		298,600	1,200		GE	
Chicago & North Western	1905		16-L-1030	6/56	LD-197	IIb	74:18	40"		298,600	1,200		GE	
Chicago & North Western	1906		16-L-1031	6/56	LD-197	IIb	74:18	40"		298,600	1,200		GE	
Chicago, St. Paul, Minneapolis & Omaha	150	CNW 150	16-L-278	2/51	LD-96	Ia	63:15	42"		287,000	800		WH	
Chicago, St. Paul, Minneapolis & Omaha	168	CNW 168	16-L-704	7/53	LD-150	Ib	63:15	42"		286,000	800		WH	
Chicago, St. Paul, Minneapolis & Omaha	169	CNW 169	16-L-705	7/53	LD-150	Ib	63:15	42"		286,000	800		WH	
Chicago, St. Paul, Minneapolis & Omaha	170	CNW 170	16-L-706	7/53	LD-150	Ib	63:15	42"		286,000	800		WH	
Chicago, St. Paul, Minneapolis & Omaha	171	CNW 171	16-L-707	7/53	LD-150	Ib	63:15	42"		286,000	800		WH	
Chicago, St. Paul, Minneapolis & Omaha	172	CNW 172	16-L-708	7/53	LD-150	Ib	63:15	42"		286,000	800		WH	
Tennessee Valley Authority	24	F3060	16-L-1157	10/58	LD-216	IIb	63:15	40"			1,200		GE	

NOTES:
1. Milwaukee Road 550-552 renumbered 527-529; 553-555 renumbered 547-549, later 524-526.

ABBREVIATIONS:
WH-Westinghouse   GE-General Electric

G. M. McDonald/*Diesel Era* collection

above, *With its horns relocated to the cab roof and an oscillating safety light added to its nose, US-TVA 24 sports a solid black paint scheme with yellow-striped pilot at Gallatin, Tenn., on July 23, 1973.* below, *Repainted in 1996, TVA F3060 carries a new road number on its cab sides, but no number in its numberboards. Gallatin, Tenn., April 13, 1996.*

J. Harlen Wilson

David R. Sweetland

Squaw Creek Coal 721001 leads a hopper train over the Yankeetown Dock Railroad near Yankeetown, Ind., in August 1970.

horsepower counterparts, many of which were retired at the expiration of their 15-year equipment leases.

One early assignment for Milwaukee's H16-66s was Madison, Wis., just 56 miles from their birthplace. Working the eastern end of the system, the six units were used in pool service with C&NW H16-66s in Michigan ore service in the late 1960s, and into the 1970s. Late in their careers, several units drifted to Kansas City for switching and transfer service.

The six units went through several renumberings. In 1959, they were re-numbered 550-555, and later, 553-555 were renumbered 547-549. The last renumbering took place when 547-552 became 524-529.

C&NW's 51-unit H16-66 fleet worked across the system before being concentrated in Upper Michigan ore service in the early 1960s. For nearly 15 years, the units could be seen working ore drags and local freights, often mated with Milwaukee's small H16-66 fleet. On September 9, 1975, Chicago & North Western retired all of its remaining H16-66s, replacing them with secondhand Norfolk & Western Alco Century 628s. ★

Train Masters in Color

Canadian National 2900 races through St. Basile-Le-Grand, Quebec, on June 30, 1962, with a commuter train.

To accommodate Canadian Pacific's request for two steam generators, CLC modified the short hood section of four Train Masters, including 8901. Calgary, Alberta, 1956.

Canadian Pacific 8901 and 8914 rest outside the Coquitlam, British Columbia, enginehouse on the outskirts of Vancouver in May 1962. Note the walkway and handrail mounted on the cab of 8901 by CP shop forces for the use of maintenance personnel.

As with the Canadian Train Masters, Jersey Central opted for a nose-mounted bell. Unit 2405 is from the carrier's first TM order, and wears the original dark green paint scheme with yellow accent stripes.

Richard Short/David R. Sweetland collection

Riding the Bethlehem, Pa., turntable, on January 10, 1969, CNJ 2406 displays the simplified paint scheme adopted by the carrier in 1965.

Robert F. Wilt/Louis A. Marre collection

Al Holtz

Arriving at Denville, N.J., in June 1955 with a four-car Train 1028, Delaware, Lackawanna & Western Train Master 982 will make a quick stop for passengers and mail before continuing on to Hoboken.

David R. Sweetland

Still wearing the paint scheme and lettering of its former owner, Erie Lackawanna 1855 waits for its next assignment in back of the Scranton (Pa.) diesel shop in the summer of 1962. When the unit was renumbered, the DL&W-style road number applied above the headlight was eliminated.

Pennsylvania Railroad H24-66s – FS24Ms in Pennsy lingo – wait at Conemaugh, Pa., in March 1961, to assist the next eastbound train over the Allegheny Mountains. Four of PRR's nine TMs were assigned to West Slope helper service.

David R. Sweetland

By the late 1960s, Pennsy concentrated its Train Master fleet in Columbus, Ohio, where the units were used in yard and transfer service. PRR 6708 sports the carrier's simplified look on September 23, 1967, at Columbus.

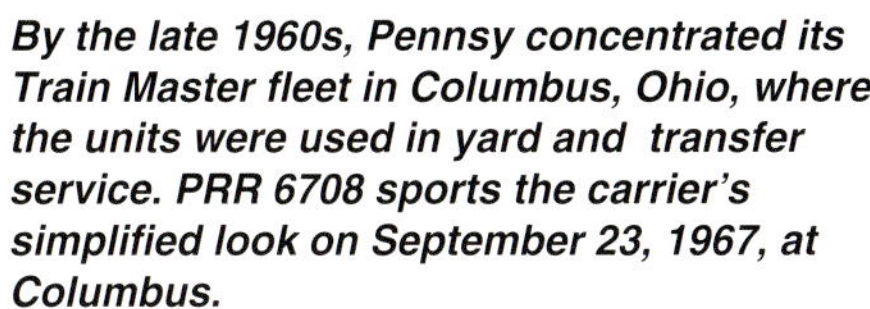

Richard O. Adams

Reading Train Master 863 leads an eastbound freight through Hershey, Pa., in August 1962. Although numbered in the passenger series, these H24-66s spent the better part of their careers hauling freight in and out of Reading's major classification yard at Rutherford, Pa.

David R. Sweetland

Mated to a GP7, Reading 802 rolls a freight through East Penn Junction, Pa., in May 1964. With an influx of second-generation diesels, the TMs spent more of their time working in hump pusher and yard service by the mid-1960s.

Robert Krone

Richard Short/David R. Sweetland collection

Idling at Erie Avenue in Philadelphia in October 1966, 867 was Reading's second and last TM painted in the yellow and green scheme. Within a year, 867 was renumbered 265; it was eventually retired in 1970 after 15 years of service.

Spending most of their time plying the rails between Cincinnati and Chattanooga, Tenn., Southern's Train Masters typically occupied the booster position between EMD F7s, and later, SD24s. Cincinnati, July 1961.

William R. Martin/David R. Sweetland collection

Alvon Thomas

Extra 4805 East *is about to leave Southern Pacific's Oakland, Calif., yard on January 31, 1959. The carman in the foreground is using a two-way radio for making the air brake test. Even after the move to California, the Train Masters were used in freight service, typically on weekends after their weekday commuter duties were over.*

David R. Sweetland

After the morning commuter rush has ended, Southern Pacific 4804 gets a complete cleaning at Mission Bay roundhouse in San Francisco on June 21, 1957.

Backing out of the station at San Francisco on June 21, 1957, SP 4812 is headed for the Mission Bay roundhouse for servicing. The Train Masters were the first road switchers to wear Espee's black widow paint scheme.

David R. Sweetland

W. L. Hammond/David R. Sweetland collection

above, *Photographed during a holiday shutdown, SP 3026 displays the scarlet and gray scheme applied to the TMs during the early 1960s. San Jose, Calif., July 2, 1972.*

Mullens, W. Va., was the hometown for Virginian's fleet of 25 Train Masters. Three of them – 60, 65, and 66 – wait at Mullens for their next run on June 1, 1958, in their attractive yellow and black scheme.

Ralph L. Phillips collection

Near DB Tower (Dickinson, W. Va.), former Virginian (now Norfolk & Western) 167 and 159 handle a short train on May 11, 1963. Although more than three years have passed since the merger, 167 still wears a faded coat of Virginian paint.

Ken Douglas/David R. Sweetland collection

Wabash 556 would soon depart for Schenectady to be repowered with an Alco 251-series engine and return with a new coat of solid blue paint and a new road number. Oakwood, Mich., January 27, 1962.

Louis A. Marre

Although the basic Train Master body was retained, Alco heavily modified the engine compartment of 593 to accommodate the new engine. Decatur, Ill., April 26, 1964.

Louis A. Marre

Former Virginian 54 has traded its colorful yellow scheme for a somber black Norfolk & Western paint job. Mullens, W. Va., September 17, 1967.

Richard O. Adams

Norfolk & Western 159, at Mullens, W. Va., on August 23, 1970, would operate another three years before being retired and rebuilt into a yard slug.

Robert F. Wilt/Louis A. Marre collection

N&W 174 is at Mullens, W. Va., on August 23, 1970. The work for which Virginian originally purchased the Train Masters changed little after Norfolk & Western acquired the carrier in 1959, and the units continued to call Mullens home.

Robert F. Wilt/Louis A. Marre collection

Milwaukee Road became the first customer for the redesigned H16-66. Featuring the railroad's distinctive frame markings, 526 is at St. Paul, Minn., on December 15, 1973.

Ed Kanak/David R. Sweetland collection

Part of Chicago & North Western's final H16-66 order, 1904 displays the Phase IIb carbody at Proviso Yard near Chicago on December 2, 1973.

Paul Hunnell/David R. Sweeetland collection

John D. Hahn, Jr.

TM-1 poses on the New York & Long Branch Railroad with a Jersey Central train at Bay Head, N.J., on September 7, 1953, with a Pennsylvania Railroad K4 class 4-6-2 serving as a backdrop.

Train Master Dispositions

RAILROAD/ OWNER	ROAD NUMBER	2nd ROAD NUMBER	3rd ROAD NUMBER	4th ROAD NUMBER	5th ROAD NUMBER	DISPOSITION
Canadian National	3000	2900				Retired Feb66
Canadian Pacific	8900					Retired 10Jun76; scrapped at Calgary (Ogden)
Canadian Pacific	8901					Retired 13Apr72; scrapped at Montreal (Angus)
Canadian Pacific	8902					Retired 11Nov65; scrapped at Calgary (Ogden)
Canadian Pacific	8903					Retired 2Apr74; sold to Preco Equipment Co., Houston, Texas, 6Nov74 via United Railway Supply, Montreal
Canadian Pacific	8904					Retired 10Jun76; scrapped at Calgary (Ogden)
Canadian Pacific	8905					Retired 10Jun76; donated to Canadian Railway Museum, Delson, Quebec
Canadian Pacific	8906					Retired 27May68; sold to Streigel Supply & Equipment
Canadian Pacific	8907					Retired 30Apr68; scrapped at Calgary (Ogden)
Canadian Pacific	8908					Retired 21Aug68; scrapped at Calgary (Ogden)
Canadian Pacific	8909					Retired 16Feb72; scrapped at Calgary (Ogden)
Canadian Pacific	8910					Retired 7Jul68; scrapped at Calgary (Ogden)
Canadian Pacific	8911					Retired 27May68; sold to Streigel Supply & Equipment
Canadian Pacific	8912					Retired 7Jul68; scrapped at Calgary (Ogden)
Canadian Pacific	8913					Retired 27May68; sold to Streigel Supply & Equipment
Canadian Pacific	8914					Retired 27May68; sold to Streigel Supply & Equipment
Canadian Pacific	8915					Retired 27May68; sold to Streigel Supply & Equipment
Canadian Pacific	8916					Retired 26Sep68; scrapped at Calgary (Ogden)
Canadian Pacific	8917					Retired 13Apr72; scrapped at Montreal (Angus)
Canadian Pacific	8918					Retired 24Oct68; scrapped at Calgary (Ogden)
Canadian Pacific	8919					Retired 27May68; sold to Streigel Supply & Equipment
Canadian Pacific	8920					Retired 24Feb69; scrapped at Calgary (Odgen)
Central Railroad of New Jersey	2401					Retired May69; sold to Naporano Iron & Metal
Central Railroad of New Jersey	2402					Retired May69; sold to Naporano Iron & Metal
Central Railroad of New Jersey	2403					Retired May69; sold to Naporano Iron & Metal
Central Railroad of New Jersey	2404					Retired May69; sold to Naporano Iron & Metal
Central Railroad of New Jersey	2405					Retired May69; sold to Naporano Iron & Metal
Central Railroad of New Jersey	2406					Retired May69; sold to Naporano Iron & Metal
Central Railroad of New Jersey	2407					Retired May69; sold to Naporano Iron & Metal
Central Railroad of New Jersey	2408					Retired May69; sold to Naporano Iron & Metal
Central Railroad of New Jersey	2409					Retired May69; sold to Naporano Iron & Metal
Central Railroad of New Jersey	2410					Retired May69; sold to Naporano Iron & Metal
Central Railroad of New Jersey	2411					Retired May69; sold to Naporano Iron & Metal
Central Railroad of New Jersey	2412					Retired May69; sold to Naporano Iron & Metal
Central Railroad of New Jersey	2413					Retired May69; sold to Naporano Iron & Metal
Delaware, Lackawanna & Western	850	EL 1850				Retired 1968; sold to Streigel Supply & Equipment Jul68
Delaware, Lackawanna & Western	851	EL 1851				Retired 1968; sold to Streigel Supply & Equipment Jul68
Delaware, Lackawanna & Western	852	EL 1852				Retired 1968; sold to Streigel Supply & Equipment Jul68
Delaware, Lackawanna & Western	853	EL 1853				Retired 1968; sold to Streigel Supply & Equipment Jul68
Delaware, Lackawanna & Western	854	EL 1854				Retired 1968; sold to Streigel Supply & Equipment Jul68

RAILROAD/ OWNER	ROAD NUMBER	2nd ROAD NUMBER	3rd ROAD NUMBER	4th ROAD NUMBER	5th ROAD NUMBER	DISPOSITION
Delaware, Lackawanna & Western	855	EL 1855				Retired 1968; sold to Streigel Supply & Equipment Jul68
Delaware, Lackawanna & Western	856	EL 1856				Retired 1968; sold to Streigel Supply & Equipment Jul68
Delaware, Lackawanna & Western	857	EL 1857				Retired 1968; sold to Streigel Supply & Equipment Jul68
Delaware, Lackawanna & Western	858	EL 1858				Retired 1968; sold to Streigel Supply & Equipment Jul68
Delaware, Lackawanna & Western	859	EL 1859				Retired 1968; sold to Streigel Supply & Equipment Jul68
Delaware, Lackawanna & Western	860	EL 1860				Retired Aug69; sold to Streigel Supply & Equipment May72
Delaware, Lackawanna & Western	861	EL 1861				Retired Oct70; sold to Streigel Supply & Equipment May72
Pennsylvania	8699	PC 6708	PC 06708			Retired 30Dec70
Pennsylvania	8700	PC 6700	PC 6799			Retired 30Dec70
Pennsylvania	8701	PC 6701	PC 06701			Retired 30Dec70
Pennsylvania	8702	PC 6702	PC 06702			Retired 30Dec70
Pennsylvania	8703	PC 6703	PC 06703			Retired 30Dec70
Pennsylvania	8704	PC 6704	PC 06704			Retired 30Dec70
Pennsylvania	8705	PC 6705	PC 06705			Retired 30Dec70
Pennsylvania	8706	PC 6706	PC 06706			Retired 30Dec70
Pennsylvania	8707	PC 6707	PC 06707			Retired 30Dec70
Reading	800					Retired 17Sep68; sold to Naporano Iron & Metal
Reading	801	201				Retired 21Apr70; sold to Luria Brothers
Reading	802					Retired 18Feb69; sold to Naporano Iron & Metal
Reading	803					Retired 18Feb69; sold to Naporano Iron & Metal
Reading	804					Destroyed in fire Jul66; retired 18Mar68; scrapped by Reading
Reading	805					Retired 18Feb69; sold to Naporano Iron & Metal
Reading	806	202				Retired 18Feb69; sold to Naporano Iron & Metal
Reading	807					Retired 24May67; sold to Streigel Supply & Equipment
Reading	808	203				Retired 21Apr70; sold to Naporano Iron & Metal
Reading	860	260				Retired 21Apr70; sold to Naporano Iron & Metal
Reading	861					Retired 7Jun70; sold to Lipsett
Reading	862	261				Retired 7Jun70; sold to Lipsett
Reading	863	262				Retired 7Jun70; sold to Lipsett
Reading	864	263				Retired 7Jun70; sold to Lipsett
Reading	865	264				Retired 7Jun70; sold to Lipsett
Reading	866					Retired 7Jun70; sold to Lipsett
Reading	867	265				Retired 7Jun70; sold to Lipsett
Southern (CNO&TP)	6300					Retired 1965; traded to EMD 23Jun65
Southern (CNO&TP)	6301					Retired 1965; traded to EMD 23Jun65
Southern (CNO&TP)	6302					Retired 1965; traded to EMD 23Jun65
Southern (CNO&TP)	6303					Retired 1965; traded to EMD 23Jun65
Southern (CNO&TP)	6304					Retired 1965; traded to EMD 23Jun65
Southern Pacific	(FM) TM-3	4800	3020			Retired Jun74; sold to Chrome Crankshaft 14Aug74
Southern Pacific	(FM) TM-4	4801	3021			Retired Dec74; sold to Chrome Crankshaft 9May75
Southern Pacific	4802	3022				Retired Dec74; sold to Chrome Crankshaft 9May75
Southern Pacific	4803	3023				Retired Jun74; sold to Chrome Crankshaft 9May75
Southern Pacific	4804	3024				Retired 1974; sold to Chrome Crankshaft 14Aug74
Southern Pacific	4805	3025				Retired Jun74 1975; to be rebuilt to brake sled; program canceled, scrapped by Levin Metals
Southern Pacific	4806	3026				Retired Jun74; sold to Chrome Crankshaft 9May75

Ken Douglas/Louis A. Marre collection

The old and the new pose at Chester, Pa., on October 5, 1963 – when they were introduced, both the Train Master and the EMD GP30 were considered to be groundbreaking locomotive designs and not just an evolutionary phase of a previous design.

As it had done on a nearly daily basis for the previous 13 years, Southern Pacific 3034 prepares to depart San Francisco in August 1970 with a San Jose-bound passenger train.

RAILROAD/ OWNER	ROAD NUMBER	2nd ROAD NUMBER	3rd ROAD NUMBER	4th ROAD NUMBER	5th ROAD NUMBER	DISPOSITION
Southern Pacific	4807	3027				Retired Dec73/Jan74; rebuilt to brake sled MW9100; scrapped by Levin Metals
Southern Pacific	4808	3028				Retired Dec73/Jan74; sold to Chrome Crankshaft 9May75
Southern Pacific	4809	3029				Retired 1974; sold to Chrome Crankshaft 14Aug74
Southern Pacific	4810	3030				Retired Dec73/Jan74; sold to Chrome Crankshaft 14Aug74
Southern Pacific	4811	3031				Retired Feb75; sold to Chrome Crankshaft 9May75
Southern Pacific	4812	3032				Retired Dec73/Jan74; sold to Chrome Crankshaft 14Aug74
Southern Pacific	4813	3033				Retired Dec73/Jan74; to be rebuilt to brake sled; program canceled, scrapped by Levin Metals
Southern Pacific	4814	3034				Retired Jun74; to be rebuilt to brake sled; program canceled, scrapped by Levin Metals
Southern Pacific	4815	3035				Retired Dec74; sold to Chrome Crankshaft 9May75
Virginian	50	N&W 150				Retired Jan70; traded to EMD
Virginian	51	N&W 151				Retired Feb67; traded to Alco
Virginian	52	N&W 152				Converted to slug Dec71, renumbered 9901; retired Nov88
Virginian	53	N&W 153	N&W 3598:3			Converted to slug Jun75, renumbered 9909
Virginian	54	N&W 154				Retired Feb70; traded to EMD
Virginian	55	N&W 155				Retired Apr71
Virginian	56	N&W 156				Retired Sep73
Virginian	57	N&W 157				Retired Sep67; traded to Alco
Virginian	58	N&W 158				Converted to slug Sep76, renumbered 9916; retired Dec92
Virginian	59	N&W 159				Converted to slug Jul74, renumbered 9904; retired Sep94
Virginian	60	N&W 160				Retired Sep67; traded to Alco
Virginian	61	N&W 161	N&W 3599:2			Converted to slug Nov76, renumbered 9917; retired Dec93
Virginian	62	N&W 162				Retired Jul70; traded to EMD
Virginian	63	N&W 163				Retired Feb70; traded to EMD
Virginian	64	N&W 164				Retired Feb70; traded to EMD
Virginian	65	N&W 165	N&W 169:2			Converted to slug Nov73, renumbered 9902
Virginian	66	N&W 166				Retired Mar67; traded to EMD
Virginian	67	N&W 167	N&W 3598:2	N&W 156:2		Converted to slug Nov74, renumbered 9908; retired Dec93
Virginian	68	N&W 168				Converted to slug Nov73, renumbered 9903; retired Dec92
Virginian	69	N&W 169	N&W 165:2			Converted to slug Dec71, renumbered 9900; retired Dec93
Virginian	70	N&W 170				Converted to slug Sep74, renumbered 9906; retired Sep94
Virginian	71	N&W 171				Converted to slug Dec75, renumbered 9913; retired Sep94
Virginian	72	N&W 172				Converted to slug Oct74, renumbered 9907; retired Dec93
Virginian	73	N&W 173				Converted to slug Sep81, renumbered 9919; retired Sep94
Virginian	74	N&W 174				Converted to slug Jul76, renumbered 9914; retired Dec92
Wabash	(FM) TM-1	550	598	N&W 3598	N&W 167:2	Converted to slug Sep74, renumbered 9905; retired Jan97
Wabash	(FM) TM-2	551	599	N&W 3599	N&W 161:2	Converted to slug Oct75, renumbered 9911; retired Sep94
Wabash	552	592	N&W 3598:4			Converted to slug Jun81, renumbered 9918; retired Jan97
Wabash	552A	595	N&W 3595	N&W 3597:2		Converted to slug Dec75, renumbered 9912; retired Jun93
Wabash	553	593	N&W 3593			Converted to slug Sep75, renumbered 9910
Wabash	553A	596	N&W 3596			Converted to slug Jul76, renumbered 9915; retired Jan97
Wabash	554	594	N&W 3594			Retired Aug70; traded to EMD
Wabash	554A	597	N&W 3597	N&W 3595:2		Retired Sep67; traded to Alco

Canadian National's lone Train Master, 2900, is at Toronto in May 1964. Note the modified end walkways to facilitate movement of crew members between the TM and other makes and models with lower end platforms.

Bibliography

Books

General

Combs, C.L., *1956 Locomotive Cyclopedia of American Practice*, New York, N.Y.: Simmons-Boardman, 1956.

Kirkland, John F. *The Diesel Builders, Volume One: Fairbanks-Morse and Lima-Hamilton*, Glendale, Calif.: Interurban Press, 1985.

Canadian Pacific

Dean, Murray W. and Hanna, David B., *Canadian Pacific Diesel Locomotives, A History of a Motive Power Revolution*, West Hill, Ontario, Canada: Railfare, 1981.

Central Railroad of New Jersey

Anderson, Elaine, *The Central Railroad of New Jersey's First 100 Years - A Historical Survey*, Easton, Pa., Center for Canal History and Technology, 1984.

Bernet, Gerard E., *Jersey Central Diesels*, Halifax, Pa.: Withers Publishing, 1990.

Delaware, Lackawanna & Western

Drury, George, *Guide to North American Steam Locomotives*, Waukesha, Wis., 1993, Kalmbach Publishing Co.

Graham, F. Stewart, *The Locomotives of the Delaware, Lackawanna & Western Railroad*, Bulletin 72, Locomotive & Railway Historical Society, Boston, July 1948.

Middleton, William D., *When the Steam Railroads Electrified*, Milwaukee: Kalmbach Publishing Co., 1974

Reading

Woodland, Dale W., *Reading Diesels, Volume 1, The First Generation*, Laury's Station, Pa.: Garrigues House, 1991.

Reading Company Technical & Historical Society, *Locomotive Diagrams of the Reading Company*, Birdsboro, Pa.: RCT&HS, 1982.

Southern Pacific

Garmany, John Bonds, *Southern Pacific Dieselization*, Edmonds, Wash.: Pacific Fast Mail, 1985.

Strapac, Joseph A. *Southern Pacific Historic Diesels, Volume 1, Fairbanks-Morse Locomotives*, Bellflower, Calif.: Shade Tree Books, 1992.

Virginian

Reid, H., *The Virginian Railway*, Milwaukee, Wis.: Kalmbach Publishing Co., 1961.

Withers, Paul K. and Bowers, Robert G., *Norfolk & Western: First Generation Diesels*, Halifax, Pa.: Withers Publishing, 1990.

Withers, Paul K. and Bowers, Robert G., *Norfolk & Western: Second Generation Diesels*, Halifax, Pa.: Withers Publishing, 1989.

Withers, Paul K. and Bowers, Robert G., *Norfolk Southern Motive Power Review 1982-1994*, Halifax, Pa.: Withers Publishing, 1995.

Books (cont.)
Wabash
Heimburger, Donald J., *Wabash*, River Forest, Ill.: Heimburger House Publishing Co., 1984.

Sweetland, David R. *Wabash in Color*, Edison, N.J.: Morning Sun Books, Inc., 1991

Periodicals
The New York Times, November 21, 1952, May 19, 1953, March 29, 1956.

Cummings, Doug., Extra 2200 South, various issues, Iron Horse Publishers, Blaine, Wash.

Pinkepank, Jerry A., *Born at Beloit*, Trains magazine, November 1964. Kalmbach Publishing Co., Milwaukee.

Ingles, J. David., *Train Master Tribute*, Trains magazine, August 1973. Kalmbach Publishing Co., Milwuakee.

Aldag, Jr., Robert. *Against the Odds-Part 1*, Trains magazine, March 1987. Kalmbach Publishing Co., Milwaukee.

________________*Against the Odds-Part 2*, Trains magazine, April 1987. Kalmbach Publishing Co., Milwaukee.

Trains magazine, October 1954, pp. 8-9

Other
Delaware, Lackawanna & Western Form 10 passenger timetable, September 27, 1953

Delaware, Lackawanna & Western Form 10A commuter timetable, February 8, 1960

Delaware, Lackawanna & Western annual report, 1953

Erie Lackawanna annual reports, 1960-1965

Fairbanks-Morse Diesel Electric Locomotives, Enginemen's Manual, Bulletin No. 1702, Fairbanks-Morse, Beloit, Wis., January 1950.

Fairbanks-Morse Diesel Electric Locomotives, Enginemen's Manual, Bulletin No. 1704, Fairbanks-Morse, Beloit, Wis., July 1950.

Fairbanks-Morse Diesel Electric Locomotives, Enginemen's Manual, Bulletin No. 1705, Fairbanks-Morse, Beloit, Wis., April 1951.

Fairbanks-Morse Diesel Electric Locomotives, Enginemen's Manual, Bulletin No. 1706, Fairbanks-Morse, Beloit, Wis., March 1952.

Fairbanks-Morse Trouble Shooters Manual-Consolidation Line Locomotives, Fairbanks-Morse, Beloit, Wis.

Jersey Central employees' timetable, October 29, 1967

Jersey Central Form TT101 timetables, April 25, 1954; April 27, 1958.

New York & Long Branch timetables March 1, 1953; April 30, 1967.

Gordon Lloyd, Jr.

Southern Pacific 3034 is at San Jose, Calif., on July 29, 1972.

With a heavy layer of sand coating the rails, Reading 863 leads Train 6, the King Coal, over Locust Summit, Pa., on April 12, 1956, while 2119, a Reading T-1 class 4-8-4, waits to cross over and return to Gordon, Pa., after pushing a freight over the hill.

J. W. Hulsman

Index

Jim Shaw

Sporting Southern Pacific's distinctive high-mounted train numberboards, 4815 rolls Train 120 through Brisbane, Calif., in 1958. SP added a plow to its Train Masters – not for snow, but for grade crossing protection.

Bruce R. Meyer

A well-worn opposed-piston engine would emit clouds of blue smoke when the throttle was rapidly advanced, as seen coming from Reading 861 as it leads an eastbound freight at speed east of Rutherford, Pa., on May 7, 1961. This view provides a good view of the large radiator cooling fans and the screened cage in which they were enclosed.